SpringerBriefs in Plant Science

SpringerBriefs present concise summaries of cutting-edge research and practical applications across a wide spectrum of fields. Featuring compact volumes of 50 to 125 pages, the series covers a range of content from professional to academic. Typical topics might include:

- A timely report of state-of-the art analytical techniques
- A bridge between new research results, as published in journal articles, and a contextual literature review
- A snapshot of a hot or emerging topic
- An in-depth case study or clinical example
- A presentation of core concepts that students must understand in order to make independent contributions

SpringerBriefs in Plant Sciences showcase emerging theory, original research, review material and practical application in plant genetics and genomics, agronomy, forestry, plant breeding and biotechnology, botany, and related fields, from a global author community. Briefs are characterized by fast, global electronic dissemination, standard publishing contracts, standardized manuscript preparation and formatting guidelines, and expedited production schedules.

Moonisa Aslam Dervash • Abrar Yousuf
Parminder Singh Sandhu • Munir Ozturk

Lantana Camara

Ecological and Bioprospecting Perspectives

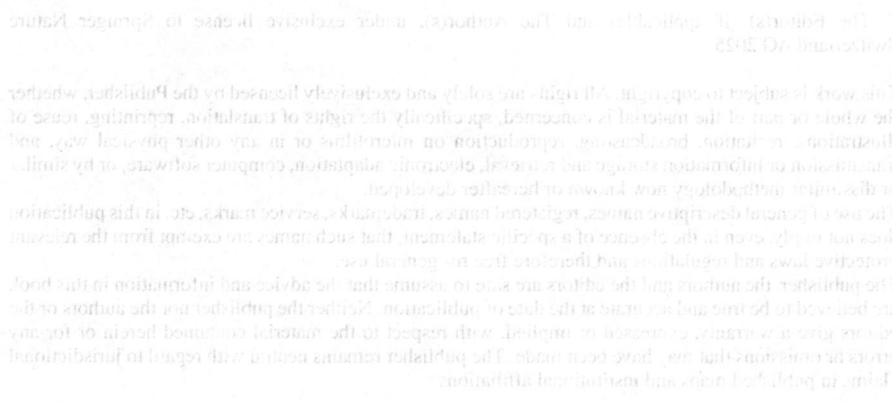

Springer

Moonisa Aslam Dervash
SKUAST-Kashmir
Srinagar, Jammu and Kashmir, India

Parminder Singh Sandhu
Punjab Agricultural University
Punjab, Punjab, India

Abrar Yousuf
Regional Research Station, Ballowal
Punjab Agricultural University
Punjab, Punjab, India

Munir Ozturk
Botany Department
Ege University
Izmir, Türkiye

ISSN 2192-1229   ISSN 2192-1210   (electronic)
SpringerBriefs in Plant Science
ISBN 978-3-032-03836-4   ISBN 978-3-032-03837-1   (eBook)
https://doi.org/10.1007/978-3-032-03837-1

This Springer imprint is published by the registered company Springer Nature Switzerland AG
The registered company address is: Gewerbestrasse 11, 6330 Cham, Switzerland

If disposing of this product, please recycle the paper.

*Dedicated to Abu Ali Husayn ibn Abd Allah
ibn Sina (Avicenna) (980–1037 CE)*

Foreword

In an age that is increasingly characterized by managing competing interests of environmental conservation and sustainable use of resources, the issue of invasive species poses both significant challenges and unanticipated benefits. *Lantana camara* is a prime example of such species with its beauty and invasive presence in tropical and subtropical regions around the world. What is most critical in presenting *Lantana camara* both as an ecologically problematic invader and as an underutilized bioprospecting opportunity is the distinctive storyline weaved throughout this valuable book.

The book, titled *Lantana camara: Ecological and Bioprospecting Perspectives*, encompasses a thorough and multifaceted perspective of *Lantana camara* beyond the typical perspective to offer a holistic idea within its expert synthesis. The authors go to great lengths to provide detailed summaries of the fundamental biology of *Lantana camara*, the invasive ecology of *Lantana camara*, and the indirect and direct consequences on the environment and socioeconomic consequences in altered environments contributing to the loss of native biodiversity, alteration of ecosystems, and consequences to agricultural producers. This thorough consideration of the problematic aspects of *Lantana camara* highlights an important base for engaging with the work.

However, this book is not just a catalogue of problems. It expertly leads the reader into the future possibilities that lurk within the shadows of the enigmatic *Lantana camara*. Through its detailed phytochemistry, the authors reveal the large variety of bioactive compounds that account for its vast range of traditional medicinal uses, cementing the connection between traditional use and science. The possible sustainable uses of *Lantana*, from environmentally friendly dyes or paper to biofuels and bioremediation are also anticipated. Regardless of the prospects of the full use of *Lantana camara*, the authors have upheld their commitment to responsible science. The chapters on *Lantana camara*'s toxicology and safety provide a significant amount of detail on understanding its risks, especially to livestock, and also the risks of products derived from *Lantana*. In the last chapter, the authors lay out the evidence behind a strong risk assessment and a robust detoxification regime.

This balanced approach is critical in uplifting any commercial use or value-added beneficial use of *Lantana camara* biomass.

In a world grappling with overexploited natural resources and the relentless march of invasive species, *Lantana camara* stands as a potent symbol of our need for innovative, integrated solutions. This book serves as an indispensable resource for researchers, environmental managers, policymakers, industry professionals, and anyone seeking a deeper understanding of invasive plant management and sustainable bioprospecting. Its balanced approach, comprehensive detail, and forward-looking perspective make it a significant contribution to the fields of ecology, phytochemistry, and natural resource management. I extend kudos to the authors for contributing this masterpiece of work to the global scientific community.

Professor of Botany & Registrar   Syed Wilayat Rizvi, M.Phil, Ph.D.
Cluster University Srinagar
Srinagar, Jammu and Kashmir, India

Preface

Lantana camara is a plant with brilliant flowers but a destructive nature and is among the most widely established invasive species in the world. Its rapid, relentless spread into many ecosystems poses threats to ecology as well as the diversity of native biodiversity. It also affects adversely the agricultural productivity and social and economic aspects of different communities. However, to view *Lantana camara* solely as an unwanted weed is to miss an important paradox: while it has a reputation as a pest, it is widely known and causing distress, possessing an unallied, rich, unexplored potential as a resource for potentially beneficial uses, a potential that is rooted in its complex and diverse phytochemistry and its historical uses. This dichotomy as the "pestilence" and "potential" in context of human interaction with *Lantana camara* is the central premise for this book.

The impetus for this compilation and realization is an immediate need for a complete understanding of *Lantana camara*. The successful management of a widespread invasive species demands more than control strategies—it requires a full understanding of its biology, ecology, effects, and importantly, its inherent value. An examination of both sides of this complex plant will pave way to arm researchers, environmental managers, policymakers, and other stakeholders with information that could turn a global environmental burden into a path for sustainable use of resources.

This book starts off by anchoring the reader in the basic issues of *Lantana camara*, giving a good biological basis with its morphology and taxonomy followed by troubling issues it presents with biological invasions. These include understanding the process of its invasion ecology and the cascade of ecological and social consequences that disrupt natural systems and humans. A section is dedicated to detailing its enormous and at times negative impacts on agriculture. In fact, it is often referred to as a blessing and a curse by farm communities.

Aside from its disruptive characteristics, we have attempted to present an organized layout discussing the surreptitious opportunity within *Lantana camara*. Following an introduction to its phytochemistry, we look into the various bioactive compounds it produces and approaches to extract those compounds. The scientific perspective directly leads to its huge history based on traditional medicine, with its evidence-based use validating forms of ancient wisdom, and revealing this plant as a truly paradoxical source of pharmacopoeia. With a good understanding of chemistry, we have tried to explore the tangible benefits *Lantana camara* can provide, including its opportunities for sustainable environmental benefits like bioremediation, and its exciting potential to develop value-added products (natural dyes and paper, biofuels, etc.).

However, we cannot understand the promise of *Lantana camara* without serious consideration of the risks it carries. The most extensive section of this book focuses on toxicology and safety, detailed enough to provide an analysis of adverse effects, specifically to livestock, but also procedures to conduct standard risk assessments and detailed pathways for detoxification. We have attempted to present safety up front to ensure responsible and ethical use.

The final sections of the volume address the balance of control and management which suggest integrative approaches combining traditional eradication approaches with innovative ones for possible use. The details presented here propose a new phase going beyond just suppression and move to a sustained symbiosis which takes *Lantana camara* from a serious adversary to an exploitable resource, with careful management.

We believe the overall approach presented here gives a rounded and authoritative account of *Lantana camara*. It will provide reliable discursive decisions for its management and innovative ideas for its sustainable use at a global benefit. This brief book will serve as a logical support for investigators with sufficient references for academicians, research scholars, and students. The researchers and connoisseurs shall find it as a comprehensive manuscript regarding *Lantana camara* fulfilling the diverse needs of readers, policymakers, teaching staff, and researchers in this field.

Srinagar, Jammu and Kashmir, India   Moonisa Aslam Dervash
Punjab, Punjab, India   Abrar Yousuf
Punjab, Punjab, India   Parminder Singh Sandhu
Izmir, Türkiye   Munir Ozturk

Acknowledgment

This manuscript has been written as a collective effort by the authors mentioned in this book. The authors are especially thankful to Dr. Kenneth Teng for his support at each and every step. Our personal gratitude to the **Springer Team** for their patience and constructive collaboration.

Contents

About the Authors

Moonisa Aslam Dervash, Ph.D., Post Doc., has obtained her Ph.D. from Division of Environmental Sciences, SKUAST-K, India. Her specialization is in environmental monitoring, environmental awareness, ecology, soil biology, wetland restoration, and carbon sequestration. She has authored 8 books with international publishers and has published more than 60 scientific research articles/book chapters in the journals/books of international reputation. She is recipient of many prestigious awards and felicitations for her dedicated accomplishments. Her focus has remained on many facets of society, especially on environmental conservation and women empowerment. She has been felicitated by state government (J&K Department of Ecology, Environment and Remote Sensing) for her outstanding contribution to mass environmental awareness and conservation through electronic media. So far, she has extended more than 10 years of dedicated service toward environmental awareness and conservation through a weekly radio program "Soun Aoundh Pukk" on All India Radio, Srinagar. She was also awarded Postdoctoral Fellowship by the Ministry of Education (ICSSR), Government of India (2022–2024). Currently, she is working in a project on "Identification of potential Rainwater Harvesting sites using Geospatial technologies in Kandi region of Punjab" funded by Science and Engineering Research Board (SERB) at Punjab Agricultural University— Ludhiana, India.

Munir Ozturk, PhD., D.Sc., received his B.Sc. (Biology-Chemistry) degree from Sri Partap College, Kashmir; M.Sc. from Postgraduate Department of Botany, Jammu & Kashmir University at Hazratbal, Kashmir; and PhD and D.Sc. from Ege University, Izmir, Turkiye. He has served at Ege University, Turkiye, for more than 50 years in different positions and has been Founder Director of the Centre for Environmental Studies, Ege University, and Chairman of the Botany Department as well as Director of the Botanical Garden. *Sideritis ozturkii* and *Verbascum ozturkii* are two newly recorded endemic plant species from Turkiye in his name. His fields of scientific interest are pollution and biomonitoring, plant eco-physiology, biosaline agriculture, and medicinal-aromatic plants conservation. He has published almost 60 books with international publishers, has authored more than 80 book chapters and 200 papers in international journals (120 with impact factor), and has presented 125 papers at international and 85 at national meetings. Dr. Munir has served as a guest editor for more than 13 international journals and holds many memberships of "Institutions and Professional Bodies." He has received fellowships from the globally recognized Alexander von Humboldt Foundation, Japanese Society for Promotion of Science, and National Science Foundation of the USA. He has worked as consultant fellow at the Faculty of Forestry, Universiti Putra Malaysia, Malaysia; as Distinguished Visiting Scientist at International Centre for Chemical and Biological Sciences, ICCBS-TWAS, Karachi University, Pakistan; as "Vice President of the Islamic World Academy of Sciences" from 2017 to 2022; and is a "Fellow of the Islamic World Academy of Science" as well as "Foreign Fellow Pakistan Academy of Science."

Parminder Singh Sandhu, PhD., District Extension Scientist (Agronomy), has completed B.Sc. Agriculture (Hons.) from Khalsa College Amritsar, Guru Nanak Dev University, Amritsar, in 2008, and M.Sc. Agronomy from Punjab Agricultural University, Ludhiana, in 2010, and Ph.D. Agronomy from Punjab Agricultural University, Ludhiana, in 2016, respectively. He is a recipient of Merit certificate in M.Sc. Agriculture, Monsanto Scholarship in M.Sc. Agriculture, Merit Scholarship in B.Sc. Agriculture (Hons), Rtn. Ajit Singh Freedom Memorial Toppers' Award by Rotary Club Amritsar in B.Sc. Agriculture (Hons), Merit certificate in B.Sc. Agriculture (Hons.), and Gold medal in B.Sc. Agriculture (Hons.). He joined PAU as Agronomist at Punjab Agricultural University—Regional Research Station, Ballowal Saunkhri, in 2016 under the scheme All India Co-ordinated Operational Research Project on Dryland Agriculture, ICAR-24 (Phase-II ORP). His field of specialization includes rainfed agriculture, dryland crops and cropping system, weed, fertilizer and irrigation management. He is involved as Co-Pl in the ICAR-sponsored project "National Innovation on Climate Resilient Agriculture (NICRA)," and "Rainfed Integrated Farming System (RIFS)." He has experience in farmer field regarding various demonstrations of dryland technologies recommended by CRIDA, Hyderabad, and PAU, Ludhiana, on farmer field. He has implemented a rainfed integrated farming system model in Kandi area of Punjab, which economically benefited farmers in various villages. Currently, involved in demonstrations of new cultivars /technologies, surveillance, monitoring and dissemination of technology, seminars, field days, workshops, campaigns, and lectures for farmers to extend latest technology among farmers. He has authored a number of extension articles, newspaper articles, and research articles in journals of national and international repute and also conducted radio talks for extension of various PAU recommendations. He is a member of many scientific associations and actively participates in scientific conferences, workshops, summer and winter schools, and training programs for building up high scientific temper and technical skills. Besides he has handled many research projects funded by RCFC-North, National Plant Medicinal Board, Ministry, AYUSH, Govt. of India, National Innovation Foundation-India, Amrapur, Gandhinagar-Mahudi Road, Gandhinagar, Gujarat and NABARD, Chandigarh.

Abrar Yousuf, Ph.D., pursued his Ph.D. from the Department of Soil and Water Engineering, Punjab Agricultural University, Ludhiana. Currently, he is working as a Scientist (Soil and Water Engineering) at Punjab Agricultural University—Regional Research Station, Ballowal Saunkhri, SBS Nagar, Punjab. His field of specialization is watershed hydrology, soil erosion modeling, watershed management, remote sensing and GIS, rainwater management, and dryland agriculture. He is involved in continuous monitoring of runoff and sediment yield from various watersheds located in Kandi region of Punjab. He has been involved in several research projects funded by various funding institutes such as ICAR New Delhi, DST New Delhi, GIZ New Delhi, IPRO Consult Germany, Department of Soil and Water Conservation, Govt. of Punjab, and SERB, New Delhi. He is working on ex situ management of rainwater in farm ponds and its judicious use through micro-irrigation systems. He has constructed farm ponds in various adopted villages. Dr. Abrar has authored a number of research articles in journals of national and international repute, technical bulletins, and two books. He is a member of several scientific associations and actively participates in scientific conferences, workshops, summer and winter schools, and training programs for building up high scientific temper and technical skills. Recently, he has been felicitated with "Best Thesis Award" for his doctoral research work by Soil Conservation Society of India, New Delhi.

Chapter 1
Introduction to *Lantana camara*:
Ecological and Bioprospecting Perspectives

Abstract This introductory chapter lays the foundations for a thorough investigation of *Lantana camara*, an invasive species that is recognized as a worldwide paradox. It results in multiple contradictions, as an invasive species of ecological concern, yet a potentially useful source for bioprospecting. It proposes how the book is designed and leads into the following chapters that discuss *Lantana camara*'s morphology and taxonomy, its rapid invasion ecology and significant environmental and social impacts, and complex agricultural affects while simultaneously introducing the history and ongoing ethnobotanical relevance of the plant, including its phytochemistry, traditional medicinal use, and potential value-added products. The chapter emphasizes perhaps the most important aspect of exploring *Lantana camara* for useful products, which is equally relevant for the health of the environment and society, toxicology, and safety. Ultimately, it establishes the book's overarching goal: that of fostering balanced understanding to convert *Lantana camara* challenges into sustainable resource use challenges and subsequent management opportunities.

1.1 The Pervasive Presence of *Lantana camara*: A Global Challenge

It is known as lantana or wild sage, *Lantana camara* is a flowering plant in the verbena family, "Verbenaceae" (Tiwari and Krishanu 2023). *Lantana camara* is native to Central and South America but is now found globally. It is predominantly found in tropical, subtropical, and warm temperate climates. Originally introduced in Europe in the seventeenth century for ornamental purposes, it became a popular garden variety due to its colorful and multicolored flowers. From Europe, *Lantana camara* spread globally through colonial trade and as an exchange from botanists (Goncalves et al. 2014). *Lantana camara* was introduced as an ornamental plant to numerous tropical and subtropical regions in the eighteenth and nineteenth centuries. However, its persistent invasiveness due to rapid growth rates, its adaptability to an array of environments, and it's out-competition of native vegetation allowed

M. A. Dervash et al., *Lantana Camara*, SpringerBriefs in Plant Science,
https://doi.org/10.1007/978-3-032-03837-1_1

this plant to become invasive. Its invasiveness is attributed to its hardiness, ability to tolerate wide ranges of soil textures, drought tolerance, and grazing by animals, birds, and other animals easily disperse its seeds.

Lantana camara has a beautiful appearance. However, *Lantana camara* poses serious ecological and economic risks in many parts of the world, where it is also quickly becoming an important pest of agriculture and ecology. It is an aggressive invasive species when introduced into an ecosystem, forming dense thickets that smother and outcompete native plants, thereby displacing them and reducing overall biodiversity and the structure and function of existing native ecosystems. It has many allelopathic characteristics, suppressing the growth of surrounding plants and establishing itself as dominant. *Lantana camara* also alters natural fire regimes, being highly flammable, by increasing the frequency and intensity of wildfires. In agricultural systems, it invades cropland and pastureland, impacting production and having a toxicity to livestock. All in all, it causes serious environmental degradation with associated economic losses (Sundaram et al. 2012).

In contrast to this negative ecological representation, there exists a rich tapestry of traditional knowledge associated with teas, ointments, and other uses and newer scientific discoveries about biological agents with substantial medical and agricultural bioprospecting potential. This book seeks to explore this complex relationship further through a detailed account of the ecological impact of this plant taxon and how this might be seen as a useful avenue for future bioprospecting activities.

1.2   Navigating the Ecological Challenge: Understanding Invasion Dynamics and Impacts

Our complete knowledge of *Lantana camara* starts with properly laying the groundwork, which is addressed more fully in Chap. 2, "Morphology and Taxonomy of *Lantana camara*: An Overview." In this chapter, we present useful information on its classification, characteristics, and reproduction that will help the reader understand how it has been so successful. Chapter 3, "Dynamics of Disruption: Invasion Ecology of *Lantana camara*," provides a much deeper review of the ecology of the plant and explains how it is able to take over rapidly, its allelochemicals and competition with native plants, and disrupting already delicate ecosystems. The invasion itself means a whole cascade of negative consequences which are described in detail in Chap. 5, "The Scrambled Web of *Lantana camara* Invasion: Ecological and Social Consequences". In this chapter, we present some of the environmental degradation and social negatives arising from the spread of this plant, including impacts on biodiversity, ecosystem services, and human livelihoods. Chapter 6 uses a narrower focus to discuss the impact of *Lantana camara* invasion on agricultural systems, even though it is associated with the "*blessing*" component that farmers deal with regularly but may rightly emphasize the "*curse*" part of the equation.

1.3 Unveiling the Bioprospecting Promise: From Traditional Uses to Modern Applications

Before evaluating *Lantana camara*'s broader implications from its destructive invasion, we turn the corner to study the plant's inherent chemical complexity. Chapter 4 entitled "Phytochemistry and Extraction of Bioactive Compounds of *Lantana camara*," explains that it is not just the mosaic of secondary metabolites (terpenoid and triterpenoid bulk/beneficial flavonoids/essential oils) but also the less illustrious methods to extract those useful compounds. The subsequent chapters present an opportunity to see the possibility beyond just being a weed.

For hundreds of years, groups of indigenous peoples have recognized and exploited some parts of this plant for its medicinal effects. Traditional Medicinal Uses and Scientific Validation of *Lantana camara* (Chap. 7) describes this great ethnomedical legacy, and the next chapter, entitled "*Lantana camara*: A Pharmacopoeia of Paradox," addresses a mixed-use in traditional healing systems, and the increased burden of scientific inquiry is confirming many of these past history claims. In addition to the medical potential, the possibility of utilizing this environmental nuisance and converting it into a resource for industries is addressed. In "Turning the Tide: *Lantana camara* for Sustainable Environmental Applications" (Chap. 9), we introduce a multitude of uses for ecological restoration, including bioremediation and biochar. The chapter "The Hidden Potential: *Lantana camara* as a Source of Value-Added Products" (Chap. 10) presents a range of industrial uses, which include natural dyes, pulp, and paper, and many bioenergy products, as examples of the uses of its biomass resource potential.

1.4 The Critical Nexus: Toxicology and Safety

It is, of course, imperative to discuss the prospects of *Lantana camara* with a full appreciation of the risks that it also entails. Chapter 11, entitled "Toxicology and Safety: Understanding and Mitigating the Risks of *Lantana camara*," will provide a critical analysis of its commonly acknowledged toxicity to animals and the potential concerns for livestock producers (primarily cattle and sheep) along with the rarer but severe risks to the human population from eating unripe berries. This important chapter includes robust risk assessment protocols and presentations of reliable and actionable detoxification plans that will allow for the use of its promising applications.

1.5 Book Structure and Objectives: A Holistic Approach

The introductory chapter presents a framework for the complete examination of *Lantana camara* presented throughout this book. The chapter outlines the principal paradox that defines the plant and provides an outline of subsequent chapters. Our goal was to integrate ecological information with opportunities for bioprospecting while prioritizing safety to give a comprehensive picture of *Lantana camara* to point the best research, policy, and practice toward making a well-known problem into opportunities for sustainable development and innovations in resource utilization. This book ends with a discussion in Chap. 12 titled "Balancing Control and Management of *Lantana camara*: Integrative Approaches and Future Directions" that suggests comprehensive research and policy options that go way beyond eradication, toward a balance of control, sustainable uses, and informed management of this fascinating taxon.

1.6 Conclusion

The introductory chapter presents a framework for the complete examination of *Lantana camara* presented throughout this book. The chapter outlines the principal paradox that defines the plant and provides an outline of subsequent chapters. Our goal was to integrate ecological information with opportunities for bioprospecting while prioritizing safety to give a comprehensive picture of *Lantana camara* to point the best research, policy, and practice toward making a well-known problem into opportunities for sustainable development and innovations in resource utilization. This book ends with a discussion that suggests comprehensive research and policy options that go way beyond eradication, toward a balance of control, sustainable uses, and informed management of this fascinating taxon.

References

Goncalves E, Herrera I, Duarte M, Bustamante RO, Lampo M, Velásquez G, Sharma GP, García-Rangel S. Global invasion of *Lantana camara*: has the climatic niche been conserved across continents? PLoS One. 2014;9:e111468. https://doi.org/10.1371/journal.pone.0111468.

Sundaram B, Krishnan S, Hiremath AJ, Gladwin J. Ecology and impacts of the invasive species, *Lantana camara*, in a social-ecological system in South India: perspectives from local knowledge. Hum Ecol. 2012;40:931–42. https://doi.org/10.1007/s10745-012-9532-1.

Tiwari P, Krishanu S. Preliminary physico—phytochemical & phyto-cognostical evaluation of the leaves of *Lantana camara*. J Pharmacogn Phytochem. 2023;12:592–6.

Chapter 2
Morphology and Taxonomy of *Lantana camara*: An Overview

Abstract *Lantana camara*, known commonly as lantana, is a tropical flowering plant native to the Americas. It is part of the Verbenaceae family and has gained notoriety as an invasive species in many places around the globe. It's a successful adaptation to various ecological niches because its morphological features make it both a pretty garden ornamental and a troublesome invader. In this chapter, the taxonomy, morphological characteristics, and how these attributes contribute to the extent of spreading by *Lantana camara* as well as its ecological impact are discussed.

2.1 Introduction

Lantana camara, usually called "lantana," is a woody, flowering shrub native to the tropics and subtropics of the Americas, particularly Central and South America. *Lantana camara* belongs to the family Verbenaceae and is taxonomically related to many aromatic and ornamental plants (Negi et al. 2019). *Lantana camara* has become a globally important flowering plant cultivated for its numerous colors, ability to adapt to climate/soil types, and overall hardiness (Munir 1996). Regardless of the horticultural benefits of *Lantana camara*, it has been recognized as a major invasive plant with 650 varieties occurring in over 60 countries, thus has emerged as an ecosystem disruptor on a global scale (Global Invasive Species Database 2020).

Lantana camara has intentionally been introduced into many countries over the past 200 years, including countries within Asia, Africa, Australia, and the Pacific Islands (Bhagwat et al. 2012). *Lantana camara* has been introduced for garden use, erosion control, wrapping or spreading, or as a natural fence (Fig. 2.1). Once *Lantana camara* has taken hold, the shrub has an exceptional ability to escape from managed landscapes to more functional and varied environments (Mungi et al. 2020). This ability to invade is primarily a combination of its morphological characteristics, reproductive potential, and ecological plasticity, which allow it to outcompete native vegetation and change the ecology of an ecosystem (Adhikari et al. 2024).

M. A. Dervash et al., *Lantana Camara*, SpringerBriefs in Plant Science,
https://doi.org/10.1007/978-3-032-03837-1_2

Fig. 2.1 *Lantana camara* bushes. (Photograph from the Kandi region of Punjab, India)

The species' ability to grow in different environmental situations, such as edges of forests to grasslands, degraded lands, as well as agricultural lands, has allowed for it to grow very dense patches which inhibit the regeneration of native plants while disrupting food chains and inhibiting habitat availability for native fauna (Ranjan 2022; Tefera et al. 2020). Additionally, *Lantana camara* has been implicated in the changing of fire regimes, soil chemistry, and grazing shapes, as well as encouraging all of these impacts from previous control attempts that increase the ecological impacts of invasive species (Berry et al. 2011; Rai 2022).

This chapter will provide an overview of the morphological and taxonomical characteristics at the species level of *Lantana camara* with implications of those biological traits that relate to spreading and overall ecological dominance (Sanders 2012). This analysis will examine the biological characteristics of this species relative to some global ecological concerns and issues, so as to emphasize the importance of understanding a plant's biology when developing appropriate management and restoration action in *Lantana* affected areas (Munir 1996).

2.2 Morphological Characteristics

Lantana camara has made a name for itself as a decorative species but considered by Global Invasive Species as one of the 100 invasive species (GISD 2022). It is also among the ten worst weeds in the world with highly detrimental ecological impacts (Lowe et al. 2000). Its invasiveness can be attributed to a suite of morphological traits that make it extremely adaptable, competitive, and difficult to remove. These traits have serious implications for it being able to invade various ecosystems globally, including tropical forests through to dry scrublands (Tefera et al. 2020). Its ability to produce large quantities of seeds that are dispersed by wildlife ensures that it can colonize new areas quickly. Furthermore, its adaptability to different

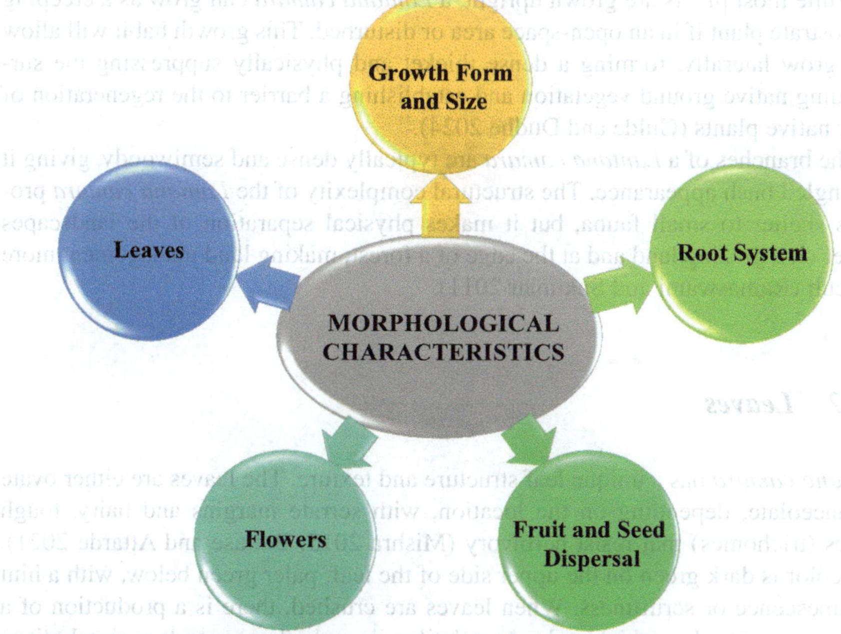

Fig. 2.2 Morphological characteristics of *Lantana camara*

environmental conditions allows it to thrive in a variety of ecosystems, including tropical forests, grasslands, and disturbed habitats such as roadsides and urban areas.

In regions where *Lantana camara* is invasive, it can outcompete native species for resources, such as light, water, and nutrients. This aggressive progression can result in the displacement of native plant species, leading to reduced biodiversity. In some areas, it has been found to alter soil chemistry, which further impacts native flora and fauna (Raphela and Duffy 2022). Additionally, its dense thickets can serve as shelter for invasive pests and other species that may disrupt local ecosystems.

In this chapter, key morphological features of *Lantana camara* shall be discussed, including growth form, leaf structure, flower structure, fruiting/seed production, and root systems (Fig. 2.2), which allow this species to invade and dominate both disturbed and undisturbed landscapes (Datta et al. 2022).

2.2.1 Growth Form and Size

Lantana camara has a flexible and hardy growth habit that adapts to a variety of environmental circumstances. It is most commonly a woody shrub, with a maximum height of 1–2 m, although it can reach up to 3 m in a good environment (Gulde and Dudhe 2024). This height allows it to compete with other vegetation around it for sunlight.

While most plants are grown upright, a *Lantana camara* can grow as a creeping or prostrate plant if in an open-space area or disturbed. This growth habit will allow it to grow laterally, forming a dense thicket and physically suppressing the surrounding native ground vegetation and establishing a barrier to the regeneration of other native plants (Gulde and Dudhe 2024).

The branches of a *Lantana camara* are typically dense and semiwoody, giving it its tangled bush appearance. The structural complexity of the *Lantana camara* provides shelter to small fauna, but it makes physical separation of the landscapes harder on a grazing land and at the edge of a forest, making land management more difficult (Ramaswami and Sukumar 2011).

2.2.2 Leaves

Lantana camara has a unique leaf structure and texture. The leaves are either ovate or lanceolate, depending on the location, with serrate margins and hairy, tough leaves (trichomes) that resist herbivory (Mishra 2015; Battase and Attarde 2021). The color is dark green on the upper side of the leaf, paler green below, with a hint of pubescence or scruffiness. When leaves are crushed, there is a production of a strong pungent odor, which is due to volatile oils and other secondary metabolites, possibly meant to reduce grazing by herbivores and also act as a pest deterrent (Mishra 2011). The leaves are arranged opposite on the stem; every node has a pair of opposite leaves on each side of the stem. This type of leaf arrangement limits competitive threat for light and can maximize light, capturing capacity photosynthesis, especially in canopy or partial shade.

2.2.3 Flowers

The flower characteristics of *Lantana camara* are important to its reproductive traits and eventual success as a nursery product. The flowers are small and tubular, borne on capitula (densely clustered and spherical inflorescences), approximately 2–3 cm in diameter (Mishra 2015). Floral colors are diverse—and include yellow, orange, red, pink, purple, and white—and frequently change as the flower develops. A flower may start as yellow and change to orange or red and subsequently may indicate to visiting insects that it has been pollinated. This ability to change color indicates to the insect visiting that the flower has been visited by another insect, thus optimizing pollinator visits and ensuring that cross-pollination efficiencies occur (Mallick and Gupta 2024). The shape of the corolla and its five-lobed opening is specialized for both bees, butterflies, and hummingbirds to effectively transfer pollen (Fig. 2.3). Floral attributes such as nectar and visual quality provide a direct floral reward to visiting pollinators, which helps to enhance their prolific reproductive output (Szigeti et al. 2018).

Fig. 2.3 Flowers of *Lantana camara*. (Photograph from the Mesopotamian region of Turkey)

2.2.4 Fruit and Seed Dispersal

After effective pollination, the *Lantana camara* plant produces fleshy, drupaceous fruit that is green before ripening to dark purple or black. Each fruit contains 1–2 hard seeds that are protected by an endocarp (Mishra 2015). The berries of *Lantana* are appealing to birds, especially frugivorous birds, which eat the fruit and defecate the seeds long distances away from the parent plant. Endozoochorous dispersal is one of the strategies that has allowed *Lantana* to quickly colonize large areas of land that are far away from its introduction site (Fukui 2005). Also, the hard seed coat allows seeds to survive for several years and contribute to a temporary seed bank in the soil. The seed bank allows for recolonization after removals are made (Hubert et al. 2024).

2.2.5 Root System

Another key morphological adaptation contributing to its invasive qualities is the root system of *Lantana camara*. It has a deep, fibrous root system, which provides good anchorage and access to deeper water and nutrient reserves (nutrient uplift), which allows the plant to tolerate drought and low soil fertility (Mishra 2015; Paudel et al. 2024). The more significant aspect of this root system is vegetative propagation, which is the ability of a plant to regenerate itself from root fragments or basal shoots. This basically means once *Lantana camara* is established, it is very hard to remove it, as the plant can reestablish itself from the root base, to regrow profusely after the aboveground biomass is removed or senesces from fire, cutting, or herbicide applications (Negi et al. 2019).

2.3 Taxonomical Characteristics

The plant *Lantana camara* was first described by the Swedish botanist Carl Linnaeus in 1753. It is classified within the genus *Lantana*, which is part of the Verbenaceae family (Munir 1996). The genus *Lantana* contains about 150 species, and *Lantana camara* is among the most widespread and invasive within this group (Negi et al. 2019).

The species *Lantana camara* has several varieties, which differ slightly in their flower colors and growth habits. These varieties have been used to distinguish different forms of the species based on morphological differences (Day et al. 2003a). Commonly, *Lantana camara* is divided into two main groups based on flower color: one with multicolored flowers and another with single-colored flowers (such as yellow, pink, red, or purple). The plant is widely distributed across tropical and subtropical regions of the world, including Asia, Africa, Australia, and the Pacific islands, as it has been introduced in many areas as an ornamental plant and for hedges (Negi et al. 2019). In tropical areas, it has also become invasive, posing a threat to local flora and fauna.

2.3.1 Taxonomy of Lantana in the Current Global Context

The complexity of taxonomy and genetics of Lantana, especially with the invasive forms such as *Lantana camara*, presents significant challenges to management in the current era of biological control and ecological management efforts. The interpretations of Sheppard (1992) and any of the subsequent researchers remain current and relevant and possibly even greater in significance, given the following globally contemporary trends:

2.3.1.1 Taxonomic Complexity and Hybridization

Lantana camara has always been taxonomically problematic with its high levels of hybridization and phenotypic plasticity (Goyal and Sharma 2015). Today, taxonomic complexities related to *Lantana camara* are heightened due to:

(i) *Globalization:* The movement of plant materials between continents has caused increased genetic mixing among *Lantana* species (Kabir et al. 2020).
(ii) *Climate Change:* Altered climatic conditions have produced *Lantana camara*'s spread to new environments, which may provide new hybrid forms (Negi et al. 2019).
(iii) *Genomic Evidence:* The use of molecular genetics and genomic sequencing is starting to reveal the hybrid nature and phylogenetic relationships of *Lantana* species, although not enough data are consistent across regions (Joshi et al. 2022).

2.3.1.2 Challenges in Biological Control

Biological control is one of the most environmentally friendly means of controlling invasive species like *Lantana camara*, but it relies on a sound understanding of our target species:

(i) *Mismatch in native and introduced ranges*: Day and Neser (2000) indicate that if biocontrol agents were collected from one of the native range *Lantana* species (e.g., Mexico, Caribbean), these agents could be less likely to adapt to hybrid populations (or variants) in the introduced range.
(ii) *Hybrid swarms*: In many countries, *Lantana camara* is represented as hybrid swarms that complicate the use of biocontrol agents due to the difficulty of matching biocontrol agents to a consistent host phenotype (Urban et al. 2011).
(iii) *Agent specificity vs adaptability*: The success of oligophagous agents (Day et al. 2003b), which could potentially develop long-term associations with the host *Lantana camara*, suggests possibilities for establishment of these species; however, their eventual efficacy and ramifications for nontarget species will need constant evaluation.

2.3.1.3 Modern Research Directions

To address these challenges, the current research and management strategies include:

(i) *Molecular Taxonomy*: DNA barcoding, genome-wide association studies (GWAS), and next-generation sequencing look to reduce species boundaries and hybrid recognition (Luo et al. 2011).

(ii) *Integrated Pest Management (IPM)*: Suitable biological control will be combined with mechanical and chemical methods for increased control effectiveness with diverse landscapes (Kishore et al. 2024).

(iii) *International Collaborations*: Shared databases and international joint fieldwork on countries impacted by *Lantana camara* can help in selecting biocontrol agents that are globally effective to control *Lantana camara* (Simelane et al. 2021).

2.3.1.4 Policy and Ecological Implications

Due to *Lantana camara*'s ecological dominance and resilience to disturbance, inaction with respect to its taxonomic complexity can have far-reaching effects on control programs globally:

(i) *Ecosystem Disruption*: *Lantana camara* has the ability to replace native flora, exploit or place limitations on fire regimes, and reduce biodiversity, all of which warrant control measures for prioritization of conservation (Shackleton et al. 2017).

(ii) *Resource Allocation*: Misidentification of *Lantana camara* forms might lead to costs in resources when selecting excess control agents, every trial and release, and monitoring (Shackleton et al. 2017).

2.4   Conclusion

Lantana camara is incredibly adaptable with characteristics that lead this plant to be at the same time persistent and invasive. Its pleasing flowers and broad ability to persist in multiple types of environmental conditions have led to its widespread distribution and replacement of natives. To effectively control their spread and reduce their ecological impacts, it is vital to understand *Lantana camara* in terms of this plant's taxon and how the morphology looks. The only way to reduce the negative effects of these invasive species on global biodiversity is through targeted research into the control and effective management.

Lantana camara's taxonomy is within the modern context globally, not just as a problematic scientific issue but as crucial to the success or failure of biological control. Coordinated global action for *Lantana camara* should bring together molecular methods, regional studies, and adaptive management required to address the issues associated with this broad, genetically variable, and biodiversity-disrupting genus.

References

Adhikari P, Lee YH, Adhikari P, Poudel A, Choi SH, Yun JY, Lee D, Park Y, Hong SH. Global invasion risk assessment of *Lantana camara*, a highly invasive weed, under future environmental change. Glob Ecol and Conserv. 2024;55:e03212. https://doi.org/10.1016/j.gecco.2024.e03212.

Battase L, Attarde D. Phytochemical and medicinal study of *Lantana camara* Linn. (Verbenaceae) – a review. Asian J Pharm Clin Res. 2021;14:20–7.

Berry ZC, Wevill K, Curran TJ. The invasive weed *Lantana camara* increases fire risk in dry rainforest by altering fuel beds. Weed Res. 2011;51:525–33. https://doi.org/10.1111/j.1365-3180.2011.00869.x.

Bhagwat SA, Breman E, Thekaekara T, Thornton TF, Willis KJ. A battle lost? Report on two centuries of invasion and management of *Lantana camara* L. in Australia, India and South Africa. PLoS One. 2012;7(3):e32407. https://doi.org/10.1371/journal.pone.0032407.

Datta DB, Das D, Sarkar B, Majumdar A. *Lantana camara* flowers as a natural dye source for cotton fabrics. J Nat Fibers. 2022;20(1):2159604. https://doi.org/10.1080/15440478.2022.2159604.

Day M, Neser S. Factors influencing the biological control of *Lantana camara* in Australia and South Africa. In: Proceedings of the X international symposium on biological control of weeds. Bozeman: Montana State University;Bozeman, Montana, USA Neal R. Spencer [ed.]. 2000. p. 897–908.

Day MD, Wiley CJ, Playford J, Zalucki MP. Will biological control of Lantana camara ever succeed? Patterns, processes & prospects. Biological Control. 2003b;28(3), 291–322. https://doi.org/10.1016/S1049-9644(03)00104-5.

Day MD, Wiley CJ, Playford J, Zalucki MP *Lantana*: current management status and future prospects. Canberra: Australian Centre for International Research, 2003a;128pp.

Fukui A. Retention time of seeds in bird guts: costs and benefits for fruiting plants and frugivorous birds. Plant Species Biol. 2005;11:141–7. https://doi.org/10.1111/j.1442-1984.1996.tb00139.x.

Global Invasive Species Database. Species profile: *Lantana camara*; 2020. Available at: https://www.iucngisd.org/gisd/species.php?sc=56

Global Invasive Species Database (GISD). 100 of the World's worst invasive alien species [online]; 2022. Available: http://www.iucngisd.org/gisd/100_worst.php

Goyal N, Sharma GP. *Lantana camara* L. (sensu lato): an enigmatic complex. NeoBiota. 2015;25:15–26. https://doi.org/10.3897/neobiota.25.8205.

Gulde S, Dudhe PS. *Lantana camara*: a medicinal plant. Int J Adv Res Sci Commun Technol. 2024;4(2):283–94. https://doi.org/10.48175/IJARSCT-22734.

Hubert B, Leprince O, Buitink J. Sleeping but not defenceless: seed dormancy and protection. J Exp Bot. 2024;75:6110–24. https://doi.org/10.1093/jxb/erae213.

Joshi AG, Praveen P, Ramakrishnan U, Sowdhamini R. Draft genome sequence of an invasive plant *Lantana camara* L. Bioinformation. 2022;18:739–41. https://doi.org/10.6026/97320630018739.

Kabir S, Kashem MA, ZabedHossain M. Genetic diversity and introductions patterns of *Lantana camara* L. in the territory of Bangladesh as revealed by microsatellite profiles. Species. 2020;21:140–9.

Kishore BSPC, Kumar A, Saikia P. Understanding the invasion potential of *Chromolaena odorata* and *Lantana camara* in the Western Ghats, India: an ecological niche modelling approach under current and future climatic scenarios. Eco Inform. 2024;79:102425. https://doi.org/10.1016/j.ecoinf.2023.102425.

Lowe S, Browne M, Boudjelas S, De Poorter M. 100 of the world's worst invasive alien species: a selection from the global invasive species database The Invasive Species Specialist Group (ISSG) a specialist group of the Species Survival Commission (SSC) of the World Conservation Union (IUCN); 2000. 12pp.

Luo L, Boerwinkle E, Xiong M. Association studies for next-generation sequencing. Genome Res. 2011;21(7):1099–108. https://doi.org/10.1101/gr.115998.110.

Mallick MAI, Gupta U. Colour selection in *Lantana camara* L. blooms: a study on butterfly attraction. Nova Geodesia. 2024;4:239. https://doi.org/10.55779/ng44236.

Mishra A. Studies of allelopathic effect of *Lantana camara* aqueous leaf extract on growth of Parthenium hysterophorus in flowering stage. Indian J Appl Res. 2011;4:33–5. https://doi.org/10.15373/2249555X/June2014/9.

Mishra A. Allelopathic properties of *Lantana camara*. Int Res J Basic Clin Stud. 2015;3:13–28. https://doi.org/10.14303/irjbcs.2014.048.

Mungi NA, Qureshi Q, Jhala YV. Expanding niche and degrading forests: key to the successful global invasion of *Lantana camara* (sensu lato). Glob Ecol Conserv. 2020;23:e01080.

Munir AA. A taxonomic review of *Lantana camara* L. and *L. montevidensis* (Spreng.) Briq. (Verbenaceae) in Australia. J Adel Bot Gard. 1996;17:1–27.

Negi GCS, Sharma S, Vishvakarma SCR, Samant SS, Maikhuri RK, Prasad RC, Palni LMS. Ecology and use of *Lantana camara* in India. Bot Rev. 2019;85:109–30. https://doi.org/10.1007/s12229-019-09209-8.

Paudel CK, Tiwari A, Baniya CB, Shrestha BB, Jha PK. High impacts of invasive weed *Lantana camara* on plant community and soil physico-chemical properties across habitat types in Central Nepal. Forests. 2024;15:1427. https://doi.org/10.3390/f15081427.

Rai PK. Environmental degradation by invasive alien plants in the Anthropocene: challenges and prospects for sustainable restoration. Archaeol Anthropol Sci. 2022;1:5–28. https://doi.org/10.1007/s44177-021-00004-y.

Ramaswami G, Sukumar R. Woody plant seedling distribution under invasive *Lantana camara* thickets in a dry-forest plot in Mudumalai, southern India. J Trop Ecol. 2011;27:365–73. https://doi.org/10.1017/S0266467411000137.

Ranjan R. Linking green bond yields to the species composition of forests for improving forest quality and sustainability. J Clean Prod. 2022;379:134708. https://ui.adsabs.harvard.edu/abs/2022JCPro.37934708R/abstract

Raphela TD, Duffy K. The impact of *Lantana camara* on invertebrates and plant species of the Groenkloof nature reserve, South Africa. Zool Stud. 2022;61:e33. https://doi.org/10.6620/ZS.2022.61-33.

Sanders RW. Taxonomy of *Lantana* Sect. Lantana (Verbenaceae): II. Taxonomic revision. Bot Res Inst Texas. 2012;6:403–41.

Shackleton RT, Witt AB, Aool W, Pratt CF. Distribution of the invasive alien weed, *Lantana camara*, and its ecological and livelihood impacts in eastern Africa. Afr J Range Forage Sci. 2017;34:1–11. https://doi.org/10.2989/10220119.2017.1301551.

Sheppard AW. Predicting biological weed control. Trends Ecol Evol. 1992;7:290–1. https://doi.org/10.1016/0169-5347(92)90224-Y.

Simelane DO, Katembo N, Mawela K. Current status of biological control of *Lantana camara* L. (sensu lato) in South Africa. Afr Entomol. 2021;29:775–83. https://doi.org/10.4001/003.029.0775.

Szigeti V, Kőrösi Á, Harnos A, Kis J. Temporal changes in floral resource availability and flower visitation in a butterfly. Arthropod Plant Interact. 2018;12:177–89. https://link.springer.com/article/10.1007/s11829-017-9585-6

Tefera N, Assefa A, Tamiru G. Distribution and abundance of invasive alien weed species in Wolayita Zone. Ethiop Cogent Food Agric. 2020;6:1778434. https://doi.org/10.1080/23311932.2020.1778434.

Urban AJ, Simelane DO, Retief F, Heystek F, Williams HE, Madire LG. The invasive 'Lantana camara L.' hybrid complex (Verbenaceae): a review of research into its identity and biological control in South Africa. Afr Entomol. 2011;19:315–48. https://doi.org/10.4001/003.019.0225.

Chapter 3
Dynamics of Disruption: Invasion Ecology of *Lantana camara*

Abstract This chapter examines the complex invasion ecology of *Lantana camara*, recognized globally as one of the most invasive plant species. The key biological and ecological characteristics that contribute to its outstanding invasive ability and impact on native ecosystems are discussed. Mechanisms examined include the high growth rate, potential for abundant seed production and seed dispersal, broad acclimatization over a large range of environmental conditions, and allelopathic ability to suppress growth of adjacent plants and vegetation. Its ability to disperse effectively and reproduce abundantly enables it to grow across diverse habitats and successfully outcompete native vegetation to cause ecological, economic, and social impacts. This chapter will explore *Lantana camara*'s invasion ecology, documenting global patterns of invasion, mechanisms of invasion, and ecological and economic consequences. In this chapter, the discussion involves the biological traits related to entering, establishing, and spreading in new locations, understanding the role of human-related activities, cover management approaches, and acknowledging actions useful to scientists, land managers, and policymakers in regions invaded by *Lantana camara*.

3.1 Introduction

In spite of its appealing ornamental appearance and bioprospecting properties, *Lantana camara* has turned into one of the most wickedly notorious invasive species in the world. Initially, introduced to various parts of the world for decorative and functional purposes, such as the ornamentals in hedges and so on—even in some cases therapeutic applications—it has spread vigorously beyond its cultivated areas into various environments, either disturbed or undisturbed (Bhodiwal et al. 2023).

Lantana camara has an aggressive dispersal ability and consequently survives in any ecosystem, thus displacing native plants and modifying ecosystems (Mhlongo et al. 2024). This plant has become a major issue for scientists, managers of land, and policymakers of invaded regions. It could grow widely over habitats and has a high reproductive capacity. This attribute makes it very destructive in dislodging native competitors, thus creating significant clear ecological, economic, and social problems (Mhlongo et al. 2024).

This chapter seeks to provide a thorough review of the invasion ecology of *Lantana camara* with emphasis on global patterns, invasion mechanisms, ecological effects, and possible management strategies (Bang et al. 2022). The biological characteristics that make it an effective intruder, anthropogenic activities, and various environmental and economic consequences of its proliferation shall be discussed (Vardien et al. 2012). Additionally, practices in its management shall be explored that have been used to control its spread and mitigate its adverse effects.

3.2 Global Distribution of *Lantana camara*

Among the plants introduced into a new region, *Lantana camara* is a prime example showing how very unconsciously human activities allow the plants to be distributed around the globe, thus leading to unforeseen ecological consequences. It had been introduced into such non-native regions during the nineteenth century. Since then, its establishment has been there in more than 60 countries, native to tropical America, extending up to almost all tropical and some subtropical areas of the world itself (Kohli et al. 2006).

3.2.1 Introduction and Spread

The spread of *Lantana camara* all over the world began in the late 1800s when people introduced *Lantana* to areas previously devoid of it as a garden and landscape ornamental plant (Kannan et al. 2013). *Lantana camara* was brought into some parts of Africa and Asia, as well as Australia, because of its horticultural trade and agricultural significance, in view of the plant being a useful hedge, erosion control, and medicinal plant.

Unfortunately, due to its very aggressive vegetative propagation, production of a huge quantity of seeds, and regeneration from root pieces quickly allowed the plant to escape from cultivation to invade natural ecosystems (Kavitha 2023; Rout et al. 2023). It rapidly invaded the tropical and subtropical regions due to the suitable climatic conditions and the ability of the plant to tolerate diverse environmental setups. In Africa, it was initially introduced for ornamental purposes and land reclamation and later emerged as an invasive species in the savannas and forest edges of this continent. Likewise, it adapted so well to disturbed habitats, like roadsides, abandoned agricultural lands, and disturbed forest patches in India and Southeast Asia (Vardien et al. 2012). It turned out that by the early twentieth century, *Lantana camara* had already caused serious concerns across the continents and regions, such as India, most parts of Southeast Asia, and Australia. It established itself in the subtropical and tropical zones of Australia, where it rapidly became one of the most common species in disturbed ecosystems, thereby displacing biodiversity and modifying native plant communities. It is also reported that this plant has spread towards some parts of South Africa, confirming its ecological adaptability with accounts of extensive infestations in the lowland and coastal regions (Goncalves et al. 2014).

1. *Africa*: Impending a vast range of ecological concerns caused by invasive plants across Africa, *Lantana camara* truly poses in possible threat across the disturbed environments of east and southern Africa. Under 2000 m altitude, it spreads fastest in logged forests and agricultural lands, altering the floristic composition; however, studies in East African savanna parks lunged toward the view that many palatable native species are almost nonexistent in *Lantana*-dominated zones (Ssali et al. 2024). Invasion corridors are demarcated by road-river interchanges and roadsides from where they climb to dominance (Ruwanza and Mhlongo 2020). Establishing dominion there greatly reduced invertebrate diversity and, therefore, the grazing potential, to the detriment of biodiversity and agricultural productivity (Samways et al. 1996; Van Wilgen et al. 2008).

2. *Southern Europe and the Middle East*: The invasion stories and ecological influences of *Lantana camara* are less documented by scientists in the regions of Southern Europe and the Middle East, where it naturalizes, for example, in parts of Spain and Portugal (Thaman 2006). It is considered an invasive species that may spread in a large number of habitats, possibly outcompeting native vegetation, whereas only a handful of very recent studies have attempted to quantify the impact of this species on biodiversity in these regions.

3. *Asia*: *Lantana camara* is highly invasive after having been introduced as an ornamental species, spreading into Asia, including India, Pakistan, Sri Lanka, and China. According to a National Survey in 2023 in India, the *Lantana camara* grew into a serious threat to biodiversity and ecosystem functioning after invading about 50% of the natural areas in the country. Research done in Mizoram, a part of the Indo-Burma biodiversity hotspot, confirms the adverse effects of *Lantana camara* invasion on native plant species diversity and density (Rai and Singh 2015). Likewise, in Sri Lanka, it travelled farther into the landscape on appeal of the British consul's wife, who was fascinated by the plant during her visit to Brazil (Guenther 1931). Bird facilitated dispersal, high reproductive capacity, and adaptability to different environmental conditions assist this rapid spread (Mungi et al. 2020; Sharma et al. 2005). Due to this invasion, the habitat of native species is reduced, altered ecosystem processes and a diminution of biodiversity oriented in the whole of Asia.

4. *Australia and New Zealand*: Introduced as an ornamental plant into Australia in 1841, *Lantana camara* escaped cultivation within two decades and now infests well over five million hectares, mostly along the eastern coast. Some investigations suggest climate change might further accelerate its spread, as habitat-suitability models project a considerable green weed range expansion in areas like the Greater Blue Mountains World Heritage Area under warmer scenarios (Kriticos et al. 2020).

5. *Island Ecosystems*: The invasive power of *Lantana camara* enjoyed a situation on many islands in the Atlantic, Pacific, and Indian Oceans. After escaping from the Royal Botanical Gardens in 1926, it rose to prominence in Sri Lanka as a weed (Ranwala et al. 2012). Introduction of the plant from Hawaii naturalized it in the Philippines. Its range, however, is presently expanding: such invasion was witnessed on islands where it was not found earlier in 1974 and included the Galapagos Islands, Saipan, and the Solomon Islands (Thaman 2006).

3.3 Factors Driving the Spread of *Lantana camara*

Human activities have had a major impact on the spread of *Lantana camara*. This plant was first introduced to various areas for its ornamental beauty, but the reasons for its spread are quite intricate. The factors responsible for the proliferation of *Lantana camara* are presented in Fig. 3.1.

The key contributors to its spread are as under:

(i) *Horticultural Trade*: As a popular ornamental plant, *Lantana camara* was extensively sold in garden centers and nurseries across the globe, especially in

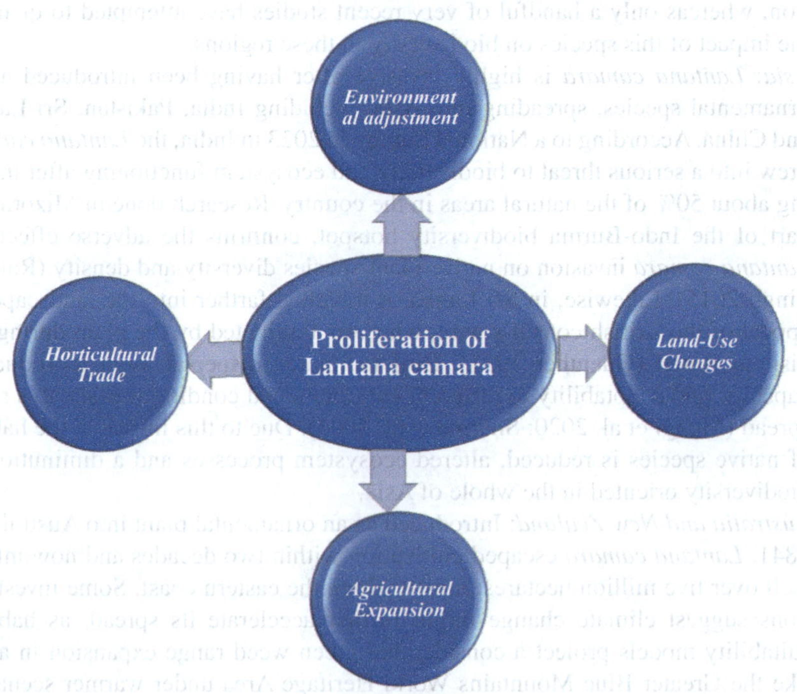

Fig. 3.1 The factors responsible for the proliferation of *Lantana camara*

tropical and subtropical regions. Unfortunately, this introduction often went unchecked, resulting in unintentional escapes from gardens (Mungi et al. 2020).

(ii) *Agricultural Expansion*: With the growth of agricultural practices worldwide, *Lantana camara* began to flourish in disturbed farmland and pastures. It quickly adapted to these new settings, outcompeting local plants and establishing itself as a troublesome weed in farming systems (Qin et al. 2016). Thus, it poses a threat to agro-ecosystems at regional, national, and global scales (Prameela et al. 2023).

(iii) *Land-Use Changes*: *Lantana camara* invades a range of disturbed habitat types that were created through land reclamation, urbanization, and deforestation. The species grows well in disturbed and novel habitats with minimal competition from other plants (such as roadsides, railroad embankments, and abandoned agricultural sites)—all types of land-use change (Arne et al. 2018).

(iv) *Environmental adjustment*: *Lantana camara* also establishes itself in invaded locations because of its high level of flexibility to different soil properties and varying climates. The excessive global spread and subsequent success of *Lantana camara* is the result of its ability to tolerate many environmental conditions, ranging from areas that are humid to those that experience drought (Goncalves et al. 2014).

3.4 Ecological Adaptations and Mechanisms of Invasion

Lantana camara can compete with native species and disperse quickly in various habitats due to its ecological adaptations and reproductive strategies, which contribute to its success as an invasive species (Adhikari et al. 2024).

3.5 High Reproductive Capacity

One of the most important factors that confers successful invasion potential is the high reproductive capacity of *Lantana camara*, which produces enormous quantities of seeds. It was also established averment that one plant can produce approximately 12,000 fruits per year (Richard et al. 2000). Mechanisms of dispersal include, but are not limited to, wind, water, and even certain animal-mediated processes (Gutierrez-Arellano and Mulligan 2018; Theoharides and Dukes 2007). Frugivorous birds (fruit-eating) have an important role in spreading seeds of *Lantana camara* to far-flung places through their droppings (Jordaan et al. 2011). It is also able to produce seeds a few months after germination, which makes it possible for rapid reproduction once established in new habitats (Theoharides and Dukes 2007).

3.5.1 Asexual Reproduction

Lantana camara, having sexual reproduction via seeds, can also regenerate vegetatively from root pieces, which makes it very hard to control. Even after cutting or mechanical removal, root parts can lead to reinstatement. This regenerative ability further increases the persistence in disturbed environments and rapid recovery following disturbances (William 2001).

3.5.2 Environmental Tolerance and Growth Rate

This flowering shrub is very tolerant of very high extremes of environmental conditions (Ramaswami and Sukumar 2013). It grows over almost all soil types, be it sandy or heavy clay, is drought-resistant, and also knows how to endure high rainfall periods. It is so fast-growing that dense thicket can supplementively be established after a short time, making it more effective in the habitats invaded. Furthermore, this species is highly tolerant to disturbed environments like roadsides, farmlands, and forest margins wherein it outcompetes the native ones (Vardien et al. 2012).

3.5.3 Allelopathy and Competitive Exclusion

Another powerful ecological weapon, allelopathy, is utilized by *Lantana camara,* in which the plant secretes inhibitory chemicals into the environment that prevent the growth of neighboring plants (Kato-Noguchi and Kurniadie 2021). Allelochemicals are the secondary metabolites composed of phenolic compounds, sesquiterpenes, triterpenes, and flavonoids. These entities have been found in the extracts, essential oil, residues, and rhizosphere soil of *Lantana camara* (Kato-Noguchi and Kurniadie 2021). These allelochemicals modify soil environment conditions in such a way that native species cannot establish or grow (Kong et al. 2006). *Lantana camara* can inhibit the growth of native grasses and herbs to a considerable extent, making it a highly competitive invasive species in invaded ecosystems. This competitive exclusion strategy rapidly invades disturbed and native communities by *Lantana camara*, accelerating the process of biodiversity loss in such areas where it is invasive (Coutts-Smith and Downey 2006).

3.6 Conclusion

Lantana camara is an extremely aggressive species with far-reaching ecological and economic consequences. Its spread across tropical and subtropical regions highlights the role of anthropogenic activities in facilitating the introduction of non-native species. While several management strategies exist to control its proliferation, *Lantana camara* remains a obstinate risk to biodiversity, agricultural productivity, and local economies.

References

Adhikari P, Lee YH, Adhikari P, Poudel A, Choi SH, Yun JY, Lee D, Park Y, Hong SH. Global invasion risk assessment of *Lantana camara*, a highly invasive weed, under future environmental change. Glob Ecol and Conserv. 2024;55:e03212. https://doi.org/10.1016/j.gecco.2024.e03212.

Arne W, Tim B, Brian WW. An assessment of the distribution and impacts of invasive alien plant species in Eastern Africa. Trans R Soc S Afr. 2018;73:217–36. https://doi.org/10.1080/003591X.2018.1529003.

Bang A, Cuthbert RN, Haubrock PJ, Fernandez RD, Moodley D, Diagne C, Turbelin AJ, Renault D, Dalu T, Courchamp F. Massive economic costs of biological invasions despite widespread knowledge gaps: a dual setback for India. Biol Invasions. 2022;24(7):2017–39. https://doi.org/10.1007/s10530-022-02780-z.

Bhodiwal S, Agarwal R, Chauhan S. A comprehensive review on various uses of *Lantana camara* L. Biospectra. 2023;18:165–70.

Coutts-Smith A, Downey P. Impact of weeds on threatened biodiversity in New South Wales. Technical report No. 11 CRC for Australian Weed Management. 2006.

Goncalves E, Herrera I, Duarte M, Bustamante RO, Lampo M, Velásquez G, Sharma GP, García-Rangel S. Global invasion of *Lantana camara*: has the climatic niche been conserved across continents? PLoS One. 2014;9(10):e111468. https://doi.org/10.1371/journal.pone.0111468. PMID: 25343481; PMCID: PMC4208836.

Gutierrez-Arellano C, Mulligan M. A review of regulation ecosystem services and disservices from faunal populations and potential impacts of agriculturalisation on their provision, globally. Nat Conserv. 2018;30:1–39. https://doi.org/10.3897/natureconservation.30.26989.

Jordaan LA, Johnson SD, Downs CT. The role of avian frugivores in germination of seeds of fleshy-fruited invasive alien plants. Biol Invasions. 2011;13:1917–30.

Kannan R, Shackleton CM, Shaanker RU. Reconstructing the history of introduction and spread of the invasive species, *Lantana*, at three spatial scales in India. Biol Invasions. 2013;15:1287–302. https://doi.org/10.1007/s10530-012-0365-z.

Kato-Noguchi H, Kurniadie D. Allelopathy of *Lantana camara* as an invasive plant. Plants. 2021;10(5):1028. https://doi.org/10.3390/plants10051028.

Kavitha CH. Allelopathic effect of aqueous leaf extracts of invasive plant species *Chromalaena odorata*, *Mimosa diplotricha* and *Combretum indicum* germination and seedling growth of some widely cultivated vegetable crops. Proceedings of a national conference, bioinvasions, trends, threats and management. Thiruvananthapuram: Kerala State Biodiversity Board; 2023. p. 137–144.

Kohli R, Batish D, Singh H, Dogra K. Status, invasiveness and environmental threats of three tropical American invasive weeds (*Parthenium hysterophorus* L., *Ageratum conyzoides* L., *Lantana camara* L.) in India. Biol Invasions. 2006;8:1501–10. https://doi.org/10.1007/s10530-005-5842-1.

Kong CH, Wang P, Zhang CX, Zhang MX, Hu F. Herbicidal potential of allelochemicals from *Lantana camara* against Eichhornia crassipes and the alga *Microcystis aeruginosa*. Weed Res. 2006;46(4):290–5.

Kriticos DJ, Clements DR, DiTommaso A. Biology of invasive plants: a new series within invasive plant science and management. Invasive Plant Sci Manag. 2020;13:115–9. https://doi.org/10.1017/inp.2020.25.

Mhlongo ES, Ruwanza S, Dalu T. Perceptions, knowledge, and invasion extent of *Lantana camara* on household yards in rural communities in Limpopo province, South Africa. Soc Nat Resour. 2024;37(8):1218–39. https://doi.org/10.1080/08941920.2024.2338773.

Mungi NA, Qureshi Q, Jhala YV. Expanding niche and degrading forests: key to the successful global invasion of *Lantana camara* (sensu lato). Glob Ecol Conserv. 2020;23:e01080.

Prameela P, Antony S, Thomas CG, Krishna VR. Plant invasions: a threat to agro-ecosystems. Proceedings of a national conference, bioinvasions, trends, threats and management. Thiruvananthapuram: Kerala State Biodiversity Board; 2023. p. 137–144.

Qin Z, Zhang JE, DiTommaso A, Wang RL, Liang KM. Predicting the potential distribution of *Lantana camara* L. under RCP scenarios using ISI-MIP models. Clim Chang. 2016;134:193–208. https://doi.org/10.1007/s10584-015-1500-5.

Rai PK, Singh MM. *Lantana camara* invasion in urban forests of an Indo–Burma hotspot region and its eco-sustainable management implication through biomonitoring of particulate matter. J Asia-Pac Biodivers. 2015;8(4):375–81. https://doi.org/10.1016/j.japb.2015.09.003.

Ramaswami G, Sukumar R. Long-term environmental correlates of invasion by *Lantana camara* (Verbenaceae) in a seasonally dry tropical forest. PLoS One. 2013;8(10):76995. https://doi.org/10.1371/journal.pone.0076995.

Ranwala S, Marambe B, Wijesundara S, Silva P, Weerakoon D, Atapattu N, Gunawardena J, Manawadu L, Gamage G. Post-entry risk assessment of invasive alien flora in Sri Lanka-present status, GAP analysis, and the most troublesome alien invaders. 23rd Asian-Pacific weed science society conference the Sebel Cairns, 26–29 September 2011; 2012.

Richard NM, Simberloff D, Lonsdale WM, Evans H, Clout M, Bazzaz FA. Biotic invasions: causes, epidemiology, global consequences, and control. Ecol Appl. 2000;10(3):689–710.

Rout Y, Naresh A, Santhosh K, Patel B. Diversity, uses and nativity of invasive alien species of Narsapur Reserve Forest, Medak, Telengana. Proceedings of national conference, bioinvasions, trends, threats and management. Thiruvananthapuram: Kerala State Biodiversity Board; 2023. p. 196–205.

Ruwanza S, Mhlongo E. *Lantana camara* invasion along road-river interchanges and roadsides in Soutpansberg, Vhembe biosphere reserve in South Africa. S Afr J Sci. 2020;116:1–5. https://doi.org/10.17159/sajs.2020/8302.

Samways MJ, Caldwell PM, Osborn R. Ground-living invertebrate assemblages in native, planted and invasive vegetation in South Africa. Agric Ecosyst Environ. 1996;59:19–32. https://doi.org/10.1016/0167-8809(96)01047-X.

Sharma GP, Raghubanshi AS, Singh JS. *Lantana* invasion: an overview. Weed Biol Manag. 2005;5:157–65. https://doi.org/10.1111/j.1445-6664.2005.00178.x.

Ssali F, Baluku R, Drileyo G, Muhumuza M. Associations between *Lantana camara* L. (Verbenaceae) and common native species in an African savanna. Ecol Solutions Evidence. 2024;5:12375. https://doi.org/10.1002/2688-8319.12375.

Thaman RR. *Lantana camara*: its introduction, dispersal and impact on islands of the tropical Pacific Ocean. Micronesia J Univ Guam. 2006;10:17–39.

Theoharides KA, Dukes JS. Plant invasion across space and time: factors affecting nonindigenous species success during four stages of invasion. New Phytol. 2007;176(2):256–73. https://doi.org/10.1111/j.1469-8137.2007.02207.x.

van Wilgen BW, Reyers B, Le Maitre DC, Richardson DM, Schonegevel L. A biome-scale assessment of the impact of invasive alien plants on ecosystem services in South Africa. J Environ Manag. 2008;89(4):336–49. https://doi.org/10.1016/j.jenvman.2007.06.015.

Vardien W, Richardson DM, Foxcroft LC, Thompson GD, Wilson JR, Le Roux JJ. Invasion dynamics of *Lantana camara* L. (Sensu lato) in South Africa. S Afr J Bot. 2012;81:81–94. https://doi.org/10.1016/j.sajb.2012.06.002.

William GL. Introduced plants, negative effects of *Lantana camara* (Verbanaceae). In: Levin SA, editor. Reference module in life sciences, encyclopedia of biodiversity. 2nd ed. Academic Press; 2001. p. 345–56. https://doi.org/10.1016/B978-0-12-384719-5.00076-9. ISBN 9780123847201.

Chapter 4
Phytochemistry and Extraction of Bioactive Compounds of *Lantana camara*

Abstract This chapter discusses the phytochemistry and extraction of bioactive components from *Lantana camara*, an invasive plant that is becoming a common plant with documented pharmacological benefits. The compounds in various parts of *Lantana camara* show notable anti-inflammatory, antimicrobial, antioxidant, antidiabetic, antitumor, and antimalarial properties. The chapter helps to describe the various classifications of bioactive molecules, extraction methods, and possible usage of the extracts as supplements, therapeutic agents, and study material for pharmaceutical applications. With proper understanding of these compounds and extraction methods, *Lantana camara* has remarkable therapeutic potential.

4.1 Introduction

Lantana camara (Linnaeus) is a member of the Verbenaceae family that serves as an invasive plant species that is present in many biodiverse ecosystems around the world and is native to the tropical parts of the Americas (Negi et al. 2019). This plant, although known for its invasive aggression in many ecosystems across the world, has gained popularity for its pharmacological capabilities. Different parts of the plant, including leaves, flowers, stem, and roots, have been researched for their bioactive compounds that have shown numerous pharmacological activities and therapeutic uses (Verma et al. 2013; Kumar et al. 2024). These therapeutic uses exhibited a variety of activities including, but not limited to, anti-inflammatory, antimicrobial, antioxidant, antidiabetic, antitumor, and antimalarial activities (Mansoori et al. 2020). As such, research on the extraction and identification of bioactive compounds from *Lantana camara* has increased in popularity (Kumar et al. 2024).

In this chapter, the bioactive compounds of *Lantana camara*, extraction methods of bioactive compounds, and potential pharmacological uses are discussed. Moreover, the categories of bioactive molecules available in the plant, detail the extraction methods, and respond to the health benefits of the bioactive compounds in the landscape of supplements, therapeutics and pharmaceuticals are also discussed.

4.2   Phytochemistry of *Lantana camara*

The phytochemical makeup of *Lantana camara* is intricate comprising a multitude of secondary metabolites with biological activity. These bioactive compounds include, but are not limited to, alkaloids, flavonoids, phenolic acids, terpenoids, glycosides, and essential oils (Mansoori et al. 2020). Each of these types of compounds contributes to the pharmacological activity of *Lantana camara*, while the diversity of the secondary metabolites can vary between the plant parts even though included in a single extraction, for example, the leaves of the plant, stems, roots, and the flowers.

4.2.1   Alkaloids

Alkaloids are pharmacologically active nitrogen-containing compounds. Numerous alkaloids from *Lantana camara*, including lantadenes, have been isolated. The most relevant use for these alkaloids is their anti-inflammatory and anticancer activity (Shah et al. 2020). The presence of alkaloids lends anticancer and immune-modulating abilities to this plant, which aids in the control of tumor cells.

4.2.2   Flavonoids

Flavonoids are a group of polyphenol compounds that are extensively known for their antioxidant, anti-inflammatory, and antidiabetic activities. Flavonoids, including quercetin, kaempferol, and luteolin, have been identified within *Lantana camara* and have considerable therapeutic value (Tiong et al. 2013). These compounds can scavenge free radicals, minimize oxidative damage, and contribute to the prevention of various chronic diseases (e.g., cardiovascular) and cancer.

4.2.3   Terpenoids

Lantana camara also has another class of bioactive compounds, terpenoids, which are reported to exert antimicrobial, anti-inflammatory and antidiabetic (González-Burgos and Gómez-Serranillos 2012). Within terpenoids, ursolic acid and oleanolic acid are among the most investigated phytochemical constituents of *Lantana camara*. The compounds are well-known for their anti-inflammatory and hepato-protective properties (Kumar et al. 2024).

4.2.4 Phenolic Acids and Other Compounds

In addition to flavonoids and terpenoids, *Lantana camara* contains phenolic acids such as chlorogenic acid, caffeic acid, and gallic acid, which contribute to the plant's antioxidant properties and are beneficial in the treatment of oxidative stress-related disorders. *Lantana camara* also includes essential oils rich in monoterpenes and sesquiterpenes, which are naturally occurring chemicals with antibacterial and insecticidal properties (Shah et al. 2020).

4.3 Extraction of Bioactive Compounds

The extraction of bioactive compounds from plant materials is typically the first step in any pharmacological research study, which is done in order to isolate and concentrate the desired molecules (Sasidharan et al. 2010). The choice of extraction method profoundly influences the yield, purity, and chemical profile of the resulting extract. Several methods have been identified to extract bioactive compounds from *Lantana camara*.

4.3.1 General Principles of Extraction

To efficiently extract compounds, a number of parameters must be optimized:

- *Solvent Selection:* This is the most important consideration. Solvents can be selected using the polarity of the desired compounds and the rule of "like dissolves like."
- *Particle Size:* Grinding the plant material increases the surface area and thus increases solvent penetration and increases extraction.
- *Temperature:* Higher temperatures typically increase solubility and diffusion rates, although they may also produce degradation of some heat vulnerable compounds.
- *Time:* There must be sufficient contact time between the solvent and the plant to extract fully.
- *Agitation:* Stirring or shaking of the mixture will improve mass transfer.

4.3.2 Traditional and Conventional Solvent Extraction Methods

These techniques are very popular because they are easy and low cost, especially at the point of the first screening or traditional preparations (Bitwell et al. 2023).

- *Decoction:* Boiling the plant material in water. Useful for water-soluble compounds like some flavonoids, some tannins, and some glycosides. Traditional medicine commonly uses this method (Pandey and Tripathi 2014).
- *Infusion:* Steeping the plant material in hot (not boiling) water. Useful for more volatile or heat-sensitive compounds. Commonly used for essential oils (but again, less efficient than distillation) and for some flavonoids (Majekodunmi 2015).
- *Maceration:* Soaking the ground plant material with a solvent (e.g., ethanol, methanol, and/or water) at room temperature for a period of time (hours to days) while occasionally stirred. A very simple way to collect a wide range of compounds (Ujang et al. 2013).
- *Percolation:* Similar to maceration but a very constant process where a solvent is allowed to flow slowly through a column filled with plant material. This method would drive off all water if kept in the percolator too long, but otherwise is more efficient at large scaling than maceration (Majekodunmi 2015).
- *Soxhlet Extraction:* Semicontinuous method that evaporates the solvent and sends the vapor to a condenser, where the cool surface allows the solvent to precipitate back to the extraction chamber. The extraction solvent is cycled repeatedly, recovering compounds into the extraction chamber. This method is particularly advantageous for compounds with reasonably good solubility in the extraction solvent, that is, nonpolar to moderately polar (e.g., triterpenoids, some components of essential oils, lipids). However, it is energy-consuming and often damaging to heat-labile compounds (Ingle et al. 2017).
- *Sequential Extraction:* A method which uses a series of solvents progressively more polar to fractionate the extract and to isolate the same groups of compounds in select fractions (Enev et al. 2021). This method is used to separate toxic nonpolar compounds (such as some lantadenes) from the more polar therapeutic compounds.

4.3.3 Advanced and Green Extraction Techniques

The benefit of these advanced methods is that they target efficiency in extraction, cutting down on the use of solvents, shorter extraction times, and less environmental impact (Bitwell et al. 2023). The various advanced and green extraction techniques are categorized as under:

- *Solvent Extraction:* Solvent extraction is regarded as the best technique. Solvent extraction is not only suitable for extracting alkaloids; other compounds can also be extracted through solvent extraction (Imane et al. 2023). As with most plants, both methanol and ethanol have been the most common and widely used solvent for extraction of *Lantana camara,* where the reported promising results from extracted phenolic compound, flavonoids and alkaloids can then be analyzed and quantified with HPLC or Gas Chromatography-Mass Spectrometry (GC-MS).
- *Steam Distillation:* Steam distillation is a method of extraction for essential oils. When steam distillation is performed, steam goes through the plant material and vaporizes the essential oils. The vapor is then condensed and collected (Zhou et al. 2023). Steam distillation is an effective method of separating volatile compounds, specifically monoterpenes and sesquiterpenes (Zhou et al. 2023), which are the main components of the essential oils of *Lantana camara.* Steam distillation is also used for extracting antimicrobial compounds of *Lantana camara* essential oils that are effective against a range of bacterial and fungal pathogens.
- *Ultrasonic-Assisted Extraction (UAE):* Is a complex system which uses ultrasonically generated waves. These waveforms create cavitation bubbles in the solvent thereby aiding the extraction process. Prior studies have demonstrated that ultrasonic-assisted extraction leads to improved yields of bioactive compounds present in plant material and will decrease the time required for extraction. Concerning bioactive compounds, UAE has been demonstrated to offer overall effectiveness in the extraction of phenolic acids and flavonoids as well as other bioactive compounds found in *Lantana camara* (Demesa et al. 2024). Additionally, UAE provides additional protection for heat-sensitive compounds as extraction temperatures are typically much lower when compared to more conventional extractions.
- *Microwave-Assisted Extraction (MAE):* Uses microwave energy to heat the solvent as well as the moisture within the plant matrix. The body then rapidly heats up and causes cell disruption and concomitantly releases compounds into the solvent. MAE allows for rapid extraction and decrease in solvent volume while also achieving high yields. MAE is efficient for flavonoids, phenolics and many triterpenoids (Altemimi et al. 2017).
- *Supercritical Fluid Extraction (SFE):* In this method, a supercritical fluid (usually carbon dioxide or CO_2) as a solvent is used. SFE has many benefits. It is much less destructive compared to methanol and is growing in popularity due to the ability to extract bioactive components from plants. SFE is also very effective for extracting components and preserving the bioactivity of sensitive components in the extract (Herzyk et al. 2024). More importantly, more than terpenoids were extracted from *Lantana camara* (González-Hernández et al. 2024). Also, SFE is a good choice for extracting lipophilic compounds. Since this method does not leave behind any toxic residues like solvents do, SFE has a smaller dependence on using toxic solvents and has a more sustainable approach in comparison to standard extraction methods (Herzyk et al. 2024).

- *Pressurized Liquid Extraction (PLE)/Accelerated Solvent Extraction (ASE):* Uses solvents at elevated temperature (up to 200 °C) and pressure (up to 20 MPa) to extract compounds. The elevated temperature dramatically increases the solubility of compounds, while elevated pressure increases mass transfer, leading to a faster, slightly more efficient procedure using less solvent compared to conventional methods. It is also versatile to extract a wide range of classes based on the selection of solvents.
- *Enzyme-Assisted Extraction (EAE):* A type of extraction that employs specifically designed enzymes (e.g., cellulases, hemicellulases, pectinases) to degrade some of the structural components of plant cell walls. The breakdown releases intracellular materials (e.g., flavonoids, phenolic compounds) into the extraction solvent, allowing for extraction with greater yields and selectivity (Puri et al. 2012).
- *Pulsed Electric Field (PEF) Extraction:* An extraction technique that uses short-duration high-voltage electrical pulses to plant material and elicit electroporation, which creates temporary pores in all membranes. Making the cell walls more permeable allows for more efficient diffusion of the materials into the solvent. This nonthermal method also protects heat-labile compounds (Ranjha et al. 2021).
- *Deep Eutectic Solvents (DES):* An emerging group of "green solvents" utilized for the extraction of a myriad of bioactive compounds. DES are made from two or more natural occurring compounds (e.g., choline chloride and urea) that based on certain ratios/treatments will form a eutectic mixture that can be liquid state at a lower temperature than the components themselves. The advantages of these solvents are the low cost, nontoxicity and environmentally friendly, and biodegradability, low melting point, and polarity (Villa et al. 2024).

4.3.4 Critical Considerations for Extraction from Lantana camara

A. *Pretreatment:* Drying (air drying, oven drying, freeze drying) and grinding materials properly to a fine powder and maximizing surface area to minimize degradation are the first and most crucial principles. Defatting (the removal of lipids or fat using nonpolar solvents e.g., hexane) can sometimes be a preliminary step (a pre-step) to improve the extraction efficiency of target compounds.
B. *Target Compound Identity:* The extraction method and solvent used will also depend on the target compounds (e.g., essential oils if working with volatiles, polar solvents for flavonoids, nonpolar for lantadenes, etc.).
C. *Toxicity Mitigation during Extraction:* There are two sides to an extraction viz.; efficient extraction of beneficial compounds and unwanted co-extraction of toxic lantadenes. Extracting target compounds while avoiding the co-extraction

of lantadenes is very challenging, which often occurs through the following steps:

(i) *Solvent Polarity Control:* Lantadenes are moderately polar. Therefore, the use of a highly polar solvent (e.g., pure water) may extract lower concentrations of lantadenes than with a polar solvent while still efficiently extracting polar flavonoids. Likewise, if very nonpolar solvents were used, the extracted essential oils and lipids will also contain some lantadenes, but likely not all of the total lantadenes that would be extracted by using the most polar solvent.

(ii) *Sequential Extraction:* Start extraction with lower polarity solvent to remove lantadenes, follow with solvents of increased polarity to extract desired therapeutic compounds.

(iii) *Fractionation:* After extraction, it is nearly always beneficial to carry out some type of purification (e.g., liquid-liquid extraction and chromatography) to fractionate the crude extract into different fractions: some enriched in the desired target compounds and depleted in toxins.

D. *Scale-up:* Scale-up from the laboratory extraction to industrial production necessitates parameter optimization with the aim of attaining the most economic, reproducible, and safe option available.

E. *Regulatory Compliance:* All extracts indicated for pharmaceutical or cosmetic use must be fabricated to the highest quality control standards, including residual solvent limits and purity profiles.

4.4 Quality Control and Characterization of Extracts

Once extraction has been achieved, it is necessary to characterize them to identify and quantify the bioactive compounds and ascertain their quality.

A. *Chromatographic Techniques:*

(i) *Thin Layer Chromatography (TLC):* A straightforward, inexpensive technique for first-attempt screening and putting a class of compound(s) into an initial category (Ingle et al. 2017).

(ii) *High-Performance Liquid Chromatography (HPLC):* Used for exact resolving, identity, and quantifying of semi-volatile or nonvolatile compounds, for example, flavonoids, triterpenoids, and phenylpropanoids. HPLC-UV and HPLC-MS are forms of HPLC (Abubakar and Haque 2020).

(iii) *Gas Chromatography-Mass Spectrometry (GC-MS):* Ideal for the identification and quantification of volatile compounds, especially the constituents of essential oils (Abubakar and Haque 2020).

B. *Spectroscopic Techniques:*

   (i) *UV-Visible Spectroscopy:* Quantification of substances in the UV-visible region of the spectrum is, of course, relevant to many phenolic compounds and flavonoids.

   (ii) *Nuclear Magnetic Resonance (NMR) Spectroscopy:* NMR is used for detailed structural analysis of isolated structures.

   (iii) *Fourier-Transform Infrared (FTIR) Spectroscopy:* In addition to confirming structure, FTIR is commonly used to identify functional groups present in the extract and also for fingerprinting.

C. *Bioassays:* In vitro and in vivo bioassays are an essential component of confirming biological activity for extracts and isolated compounds and correlating chemical constituents of extracts with pharmacological effects (e.g., antioxidant, antimicrobial, and anti-inflammatory assays).

4.5   Pharmacological Applications of Bioactive Compounds from *Lantana camara*

Medical properties of bioactive compounds from *Lantana camara* are well-studied. These properties include pharmacological activities such as anti-inflammatory, antimicrobial, antioxidant, antidiabetic, and anticancer responses (Barros et al. 2017).

4.5.1   *Antioxidant Activity*

Lantana camara has potential antioxidant activity due to the presence of many flavonoids, phenolic acids, and terpenoids. The extracts and fractions of *Lantana camara* have strong free radical scavenging ability, particularly in preventing oxidative damage related to chronic diseases like cancer, cardiovascular diseases, and neurodegenerative diseases (Anand et al. 2018).

4.5.2   *Antimicrobial and Antifungal Properties*

Lantana camara, as an essential oil and extract, has shown to have antimicrobial activity. *Lantana camara* has been evaluated for antimicrobial activity against a variety of pathogens (*E. coli* and *S. aureus* bacteria; *Candida albicans* fungi). *Lantana camara* could be a candidate for natural antimicrobial activity (Ruburika et al. 2022).

4.5.3   Anti-inflammatory and Anticancer Activity

Some studies demonstrated that some compounds, such as ursolic acid and oleanolic acid in *Lantana camara,* exert potent anti-inflammatory activity by blocking the production of pro-inflammatory cytokines. Laboratory studies regarding cytotoxicity and anti-inflammatory aspects concluded that *Lantana camara* extract was at least safer than piroxicam while showing strong anti-inflammatory activity (El-Banna et al. 2022). In addition, *Lantana camara* extracts have shown promise in cancerous cell proliferation inhibiting benefits and studies indicate this could be useful as adjunct therapies for cancer (Bajpai et al. 2024).

4.5.4   Antidiabetic Activity

The studies regarding the anti-diabetic potential of *Lantana camara* have provided evidence to support that some bioactive compounds from the plant may contribute collectively and/or antagonistically to maintaining blood glucose levels. Most importantly, some of the flavonoids, including quercetin, have been directly related to the ability of the plant to reduce blood glucose and improve insulin sensitivity (Salehi et al. 2019).

4.6   Conclusion

Lantana camara is a plant species identified with a variety of phytochemicals and offers numerous bioactive substances with possible pharmacological applications. The bioactive constituents of the plants are known to have alkaloids, flavonoids, terpenoids, and phenolic acids as therapeutic properties (antioxidant, antimicrobial, anticancer, antidiabetic, etc.). New extraction methods (e.g., solvent extraction, steam distillation, supercritical fluid extraction, and ultrasound-assisted extraction) are ensuring less cumbersome processes.

Further studies are needed to assess full clinical applicability, but *Lantana camara* has opportunities as a natural product source towards the development of new drugs and therapeutics. Science will continue to focus on exploring the bioactive components of the plants, using better extraction methods, which could be applicable to alternate sustainable approaches in ensuring solutions to contemporary health solutions.

References

Abubakar AR, Haque M. Preparation of medicinal plants: basic extraction and fractionation procedures for experimental purposes. J Pharm Bioallied Sci. 2020;12(1):1–10. https://doi.org/10.4103/jpbs.JPBS_175_19.

Altemimi A, Lakhssassi N, Baharlouei A, Watson DG, Lightfoot DA. Phytochemicals: extraction, isolation, and identification of bioactive compounds from plant extracts. Plants. 2017;6(4):42.

Anand J, Chaudhary S, Rai N. Analysis of antioxidant activity, total phenolic content and total flavonoid content of *Lantana camara* leaves and flowers. Asian J Pharm Clin Res. 2018;11:203. https://doi.org/10.22159/ajpcr.2018.v11i4.23900.

Bajpai P, Usmani S, Kumar R, Prakash O. Recent advances in anticancer approach of traditional medicinal plants: a novel strategy for cancer chemotherapy. Intell Pharm. 2024;2(3):291–304. https://doi.org/10.1016/j.ipha.2024.02.001.

Barros LM, Duarte AE, Pansera Waczuk E, Roversi K, da Cunha FAB, Rolon M, Coronel C, Gomez MCV, de Menezes IRA, da Costa JGM, Boligon AA, Hassan W, Souza DO, da Rocha JBT, Kamdem JP. Safety assessment and antioxidant activity of *Lantana montevidensis* leaves: contribution to its phytochemical and pharmacological activity. EXCLI J. 2017;16:566–82. https://doi.org/10.17179/excli2017-163.

Bitwell C, Indra SS, Luke C, Kakoma MK. A review of modern and conventional extraction techniques and their applications for extracting phytochemicals from plants. Sci Afr. 2023;19:01585. https://doi.org/10.1016/j.sciaf.2023.e01585.

Demesa AG, Saavala S, Pöysä M, Koiranen T. Overview and toxicity assessment of ultrasound-assisted extraction of natural ingredients from plants. Foods. 2024;13(19):3066. https://doi.org/10.3390/foods13193066.

El-Banna AA, Darwish RS, Ghareeb DA, Yassin AM, Abdulmalek SA, Dawood HM. Metabolic profiling of *Lantana camara* L. using UPLC-MS/MS and revealing its inflammation-related targets using network pharmacology-based and molecular docking analyses. Sci Rep. 2022;12(1):14828. https://doi.org/10.1038/s41598-022-19137-0.

Enev V, Sedláček P, Kubíková L, Sovová Š, Doskočil L, Klučáková M, Pekař M. Polarity-Based sequential extraction as a simple tool to reveal the structural complexity of humic acids. Agronomy. 2021;11(3):587. https://doi.org/10.3390/agronomy11030587.

González-Burgos E, Gómez-Serranillos MP. Terpene compounds in nature: a review of their potential antioxidant activity. Curr Med Chem. 2012;19(31):5319-41. https://doi.org/10.2174/092986712803833335. PMID: 22963623.

González-Hernández RA, Valdez-Cruz NA, Trujillo-Roldán MA. Factors that influence the extraction methods of terpenes from natural sources. Chem Pap. 2024;78:2783–810. https://doi.org/10.1007/s11696-024-03339-z.

Herzyk F, Piłakowska-Pietras D, Korzeniowska M. Supercritical extraction techniques for obtaining biologically active substances from a variety of plant byproducts. Foods. 2024;13(11):1713. https://doi.org/10.3390/foods13111713.

Imane G, Hemmami H, Ben Amor I, Zeghoud S, Ben Seghir B, Hammoudi R. Different methods of extraction of bioactive compounds and their effect on biological activity: a review. Int J Secondary Metab. 2023;10:469–94. https://doi.org/10.21448/ijsm.1225936.

Ingle KP, Deshmukh AG, Padole DA, Dudhare MS, Moharil MP, Khelurkar VC. Phytochemicals: extraction methods, identification and detection of bioactive compounds from plant extracts. J Pharmacogn Phytochem. 2017;6(1):32–6.

Kumar R, Guleria N, Deeksha MG, Kumari N, Kumar R, Jha AK, Parmar N, Ganguly P, de Aguiar Andrade EH, Ferreira OO, de Oliveira MS, Chandini. From an invasive weed to an insecticidal agent: exploring the potential of *Lantana camara* in insect management strategies – a review. Int J Mol Sci. 2024;25:12788. https://doi.org/10.3390/ijms252312788.

Majekodunmi SO. Review of extraction of medicinal plants for pharmaceutical research. Merit Res J Med Med Sci. 2015;3:521–7.

Mansoori A, Singh N, Dubey SK, Thakur TK, Alkan N, Das SN, Kumar A. Phytochemical characterization and assessment of crude extracts from *Lantana camara* L. for antioxidant and antimicrobial activity. Front Agron. 2020;2:582268. https://doi.org/10.3389/fagro.2020.582268.

Negi GCS, Sharma S, Vishvakarma SCR, Samant SS, Maikhuri RK, Prasad RC, Palni LMS. Ecology and use of *Lantana camara* in India. Bot Rev. 2019;85:109–30. https://doi.org/10.1007/s12229-019-09209-8.

Pandey A, Tripathi S. Concept of standardization, extraction and pre-phytochemical screening strategies for herbal drug. J Pharmacog Phytochem. 2014;2(5):115–9.

Puri M, Sharma D, Barrow CJ. Enzyme-assisted extraction of bio-actives from plants. Trends Biotechnol. 2012;30(1):37–44. https://doi.org/10.1016/j.tibtech.2011.06.014.

Ranjha M, Kanwal R, Shafique B, Arshad RN, Irfan S, Kieliszek M, Kowalczewski PŁ, Irfan M, Khalid MZ, Roobab U, Aadil RM. A critical review on pulsed electric field: a novel technology for the extraction of phytoconstituents. Molecules. 2021;26(16):4893. https://doi.org/10.3390/molecules26164893.

Ruburika IN, Izere C, Habyarimana T, Niyonzima FN. Antimicrobial activity of *Lantana camara* extracts against selected clinical isolated bacteria. Int J Health Sci. 2022;6(S3):10828–37. https://doi.org/10.53730/ijhs.v6nS3.8437.

Salehi B, Ata A, Kumar NAV, Sharopov F, et al. Antidiabetic potential of medicinal plants and their active components. Biomolecules. 2019;9(10):551. https://doi.org/10.3390/biom9100551.

Sasidharan S, Chen Y, Saravanan D, Sundram KM, Yoga Latha L. Extraction, isolation and characterization of bioactive compounds from plants' extracts. Afr J Tradit Complement Altern Med. 2010;8(1):1–10.

Shah MM, Alharby H, Hakeem K. *Lantana camara*: a comprehensive review on phytochemistry, ethnopharmacology and essential oil composition. Lett Appl NanoBioSci. 2020;9:1199–207. https://doi.org/10.33263/LIANBS93.11991207.

Tiong SH, Looi CY, Hazni H, Arya A, Paydar M, Wong WF, Cheah S, Mustafa MR, Awang K. Antidiabetic and antioxidant properties of alkaloids from *Catharanthus roseus* (L.) G. Don Mol. 2013;18:9770–84. https://doi.org/10.3390/molecules18089770.

Ujang ZB, Subramaniam T, Diah MM, Wahid HB, Abdullah BB, Abd-Rashid AHB, Appleton D. Bio-guided fractionation and purification of natural bio-actives obtained *from Alpinia conchigera* water extract with melanin inhibition activity. J Biomater Nanobiotechnol. 2013;4(3):265–72.

Verma SC, Jain CL, Nigam S, Padhi MM. Rapid extraction, isolation, and quantification of oleanolic acid from *Lantana camara* L. roots using microwave and HPLC-PDA techniques. Acta Chromatogr. 2013;25:181–99. https://doi.org/10.1556/achrom.25.2013.1.12.

Villa C, Caviglia D, Robustelli della Cuna FS, Zuccari G, Russo E. NaDES application in cosmetic and pharmaceutical fields: an overview. Gels. 2024;10(2):107. https://doi.org/10.3390/gels10020107.

Zhou W, Li J, Wang X, Liu L, Li Y, Song R, Zhang M, Li X. Research progress on extraction, separation, and purification methods of plant essential oils. Separations. 2023;10(12):596. https://doi.org/10.3390/separations10120596.

Chapter 5
The Scrambled Web of *Lantana camara* Invasion: Ecological and Social Consequences

Abstract This chapter discusses the significant ecological and sociodemographic outcomes of the global invasion of *Lantana camara* in the form of the scrambled web. It is regarded as a highly flexible and adaptable species, occupying a variety of ecosystems, in which it quickly transforms these into *Lantana*-invaded ecosystems, with consequences as far-reaching as losses in biodiversity, degradation of habitat, and changes to ecosystem functions in regard to fire regime. Its invasion was and continues to have socioeconomic impacts, including extensive degradation of farmland, production losses in the agricultural and horticultural sector, and overall livelihood impacts. The chapter will unpack and elucidate the complexities of managing this persistent invader.

5.1 Introduction

For centuries, this unremarkable *Lantana camara* has been orchestrated to invade the terrestrial ecosystems across the globe. Native to the Americas, this species has demonstrated a stunning ability to populate a plethora of environments and aggressively extend its already vast range across distant continents and exotic biodiversity landscapes in Asia and Africa, into the exotic ecology of Australia, and the remote, often fragile environments of the Pacific Islands (Lone et al. 2025). Such relentless groping for new real estate has very much made *Lantana camara* into the world-over-recognized invasive species; this is a stark example of how an introduced species with good intentions can be, somehow, inflicted with calamity upon human existence (Negi et al. 2019).

Originally valued for its bright and varied floral brilliance in gardens, in land restoration, and in indigenous medicine, the very nature of *Lantana camara* as an adaptable and opportunistic invader soon came to be seen (Patel et al. 2010). It pounds relentlessly and brings about far-reaching ecological and socioeconomic disruptions in its new habitats. This chapter, therefore, delves into some of these impacts occurring far and wide. A thorough look at the cascading impacts of

M. A. Dervash et al., *Lantana Camara*, SpringerBriefs in Plant Science,
https://doi.org/10.1007/978-3-032-03837-1_5

Lantana camara on the native biodiversity, describing its competition to native flora and fauna and the consequent degradation of habitats and declines in native flora and fauna, shall be reviewed. It will also be discussed in great detail how *Lantana* affects the basic functions of ecosystems, changes in nutrient cycling, water availability, and fire regimes. This will include explanations of the degradation of farmlands and the consequent reduction in crop yields as the invasive *Lantana camara* species encroaches upon formerly productive agricultural lands. Much attention will be given to the implications for local communities, with a focus on how the invasive species impacts livelihoods, traditional practices, and human well-being. Thus, the chapter will venture into the complex problems posed by managing such a widely spread and stubborn plant and will analyze the wider social and economic repercussions of its continued global spread so as to stress the necessity of control and awareness programmes.

5.2   Ecological Consequences

5.2.1   *Impacts on Biodiversity*

The foremost ecological impact of *Lantana camara* invasion is to displace native plant species and reduce biodiversity (David et al. 2017). Aggressive growth, competition for resources, and unpalatability make the plant a competitor of native flora, rapidly colonizing the invaded and uninvaded habitats (Abebe 2018). It has been shown to form dense, monoculture thickets that reduce the abundance and diversity of native plant species. Invasive species like *Lantana camara* cast a negative impact on invertebrates and native plants. This transformation in plant composition has cascading effects on herbivores, pollinators, and other organisms that rely on native plants for food and habitat thereby paralysing food chains and webs (Raphela and Duffy 2022). Thus, this noxious weed affects ecosystem dynamics and nutrient flows in the infested landscapes (Argaw 2016).

In tropical and subtropical ecosystems where the usual condition is high plant diversity, *Lantana camara* has become a serious danger to native ecosystems (Tadele 2014). In Australia, the plant is a significant invader of rainforests, woodlands, and grasslands; it changes the structure of plant communities, thereby reducing the food and shelter available for native fauna (De Lacy and Shackleton 2017). Likewise, in some parts of India and Southeast Asia, *Lantana camara* has displaced the native flora, mainly in forest edges and in degraded areas, leading to further loss of species diversity and habitat heterogeneity (Chaudhary et al. 2021).

5.2.2   Soil and Habitat Alterations

In addition to exerting its direct effects on native flora, *Lantana camara* also brings about alteration in the habitat's physical structure (Gooden et al. 2009). The light penetration is restricted due to the thickened canopy made by the thickets of this species, thereby suppressing herbaceous and small shrubby growth. Thus, the modification of the understory can cause long-term repercussions on forest regeneration (Sharma and Raghubanshi 2007). In tropical forests, where *Lantana camara* is the most aggressive invasive, suppression of native vegetation leads to a modification in forest structure and, consequently, can change ecosystem functioning, such as nutrient cycling and carbon storage (Lone et al. 2025).

In addition, *Lantana camara* changes soil chemistry and structure. Since, *Lantana camara* is allelopathic, obtaining its defensive influences on germination and early growth of plant species (Alemayehu 2024), the chemicals are released in the soil by decaying leaves and roots, thus preventing the establishment of native species (Mack et al. 2000). Allelopathy adds to the competitive edge of *Lantana camara*, strengthening its position in compromised sites and inhibiting the recovery of native plant communities.

5.2.3   Disruption of Pollination and Seed Dispersal Networks

Lantana camara invasion disrupts native pollination and seed dispersal systems. While *Lantana camara* is capable of attracting various pollinators, including bees, butterflies, as well as birds, its presence can engender shifts in the dynamics of the native pollination systems. When the flowers of *Lantana camara* grow overwhelmingly dominant, native pollinators may be displaced, especially in those areas where native plants require specific pollinators for their reproduction (Carrión-Tacuri et al. 2014). This interference can thus further hamper the reproductive capacity of indigenous plants, thereby compounding biodiversity loss.

Seed dispersal of native plants is also affected by *Lantana camara* because dispersed by birds and mammals, and by wind as well, the seed can compete with native plants for dispersal mechanisms (Sundaram et al. 2012). In some areas, seeds of *Lantana camara* get dispersed at a great distance from their origin, thereby enabling the species to establish itself at further sites and invade undisturbed habitats (Mungi et al. 2020).

5.3   Social and Economic Consequences

The proliferation of *Lantana camara* has far-reaching ecological and economic impacts. It poses a direct threat to biodiversity, agricultural productivity, and local economies in regions, where it has become an invader.

5.3.1 Impacts on Agriculture

Lantana camara is a major pest in Agriculture, especially in tropical and subtropical regions. The plant invades pasturelands, crop fields, and other agricultural landscapes, thus presenting a threat to food production and farming as a livelihood (Ntalo et al. 2022). The plant crowds out other vegetation with thick thickets, reduces grazing land for livestock, and competes with crops for water and nutrients. This directly impacts the farmers since forage quality and quantity are reduced by *Lantana camara*, especially in areas where vegetables or crop pasture lands are kept for livestock grazing (Shackleton et al. 2017).

In regard to soil fertility, the developments of *Lantana camara* have been associated with its depletion. Rapid growth and dense canopy cast heavy shade, obstructing the deposition of sunlight and nutrients to the surrounding vegetation, which, once inappropriately dashed, may bring about degradation of the soil of that locale (Negi et al. 2019). This, in turn, is another vicious circle, as *Lantana camara* degrades the soil so much further, thus inhibiting the growth of native plants and crops.

Furthermore, *Lantana camara* is toxic to livestock, particularly cattle and horses. Its toxicity is due to various secondary metabolites (naphthoquinones, iridoids, glycosides, and the most lethal lantadenes). These lantadenes are actually pentacyclic triterpenes that interfere with the liver functioning, resulting in hepatotoxicity and photosensitization in grazing animals (Sharma et al. 2007). Consumption of any part of the plant leads to poisoning, decreased productivity of animals, and even death in some cases (Sharma and Makkar 1981), with goats being least affected than other grazers due to higher metabolic tolerance (Lal and Kalra 1960). This definitely entails adverse consequences on the economic situation of farmers, more so in places where livestock forms an essential part of agricultural economy (Sharma et al. 1981).

5.3.2 Impact on Human Health and Well-Being

Besides its economic consequences, this plant can also be dangerous to human health in some cases, as the toxic substances contained in its leaves, berries, and flowers can cause skin irritation and gastrointestinal troubles if ingested (Nawaz et al. 2016). There have also been reports of *Lantana camara* poisoning in children after eating an unripe fruit (Wolfson and Solomons 1964). The spread of the plant into rural and urban areas may damage the ecosystems from which people carve out a living or collect wood, medicinal plants, and food. As *Lantana camara* invades areas, it threatens these vital resources and hence the livelihood of rural communities.

5.3.3 Social and Cultural Impacts

In many regions around the world where *Lantana camara* has invaded, local landscapes and cultural practices have been transformed by it (William 2001). In the Indian subcontinent, *Lantana camara* has encroached upon sacred groves, which become significant from the perspective of indigenous communities. The overbearing presence of the plant interferes with local cultural practices that revolve around traditional land use, ceremonies, and the gathering of medicinal plants. *Lantana camara* incursion into sacred and traditionally important places thus also brings about the erosion of cultural heritage and practices so deeply intertwined with the natural world (Negi et al. 2019).

From the agro-related viewpoints, *Lantana camara* is termed as the plant that makes management of lands in some areas tough for communities, depending on agriculture and natural resources for their livelihood paving way to certain social and economic tensions, as the population in those areas struggles to come to terms with the loss of productive land and access to resources (Negi et al. 2019).

5.4 Conclusion

Lantana camara is a highly invasive species with severe ecological, agricultural, and social consequences. It disperses itself through various native ecosystems, taking opportunities for the native species, lessening biodiversity, damaging agricultural productivity, and, therefore, bringing in economic losses, with food security under threat. It presents a grave concern in the rangelands and pastures where, in turn, it becomes a hindrance for its surrounding communities.

References

Alemayehu Y, Chimdesa M, Yusuf Z. Allelopathic Effects of Lantana camara L. Leaf Aqueous Extracts on Germination and Seedling Growth of Capsicum annuum L. and Daucus carota L. Scientifica (Cairo). 2024 Apr 16;2024:9557081. https://doi.org/10.1155/2024/9557081. PMID: 38962531; PMCID: PMC11221968.

Abebe FB. Invasive *Lantana camara* L ahrub in Ethiopia: ecology, threat, and suggested management strategies. 2018;10(7):184–95. https://doi.org/10.5539/jas.v10n7p184.

Argaw A. Biodegradation of *Lantana camara* using different animal manures and assessing its manurial value for organic farming. Saudi J Biol Sci. 2016;2(1):52–60. https://doi.org/10.22205/sijbs/2016/v2/i1/100344.

Carrión-Tacuri J, Berjano R, Guerrero G, Figueroa E, Tye A, Castillo J. Fruit set and the diurnal pollinators of the invasive *Lantana camara* and the endemic *Lantana peduncularis* in the Galapagos Islands. Weed Biol Manag. 2014;14:209–2019. https://doi.org/10.1111/wbm.12048.

Chaudhary A, Sarkar MS, Adhikari BS, Rawat GS. *Ageratina adenophora* and *Lantana camara* in Kailash sacred landscape, India: current distribution and future climatic scenarios through modeling. PLoS One. 2021;16(5):0239690. https://doi.org/10.1371/journal.pone.0239690.

David P, Thebault E, Anneville O, Duyck PF, Chapuis E, Loeuille N. Impacts of invasive species on food webs: a review of empirical data. Adv Ecol Res. 2017;56:1–60. https://doi.org/10.1016/bs.aecr.2016.10.001.

De Lacy P, Shackleton CM. Woody plant species richness, composition and structure in urban sacred sites, Grahamstown, South Africa. Urban Ecosyst. 2017;20:1169–79. https://doi.org/10.1007/s11252-017-0669-y.

Gooden B, French K, Turner PJ, Downey PO. Impact threshold for an alien plant invader, *Lantana camara* L., on native plant communities. Biol Conserv. 2009;142:2631–41. https://hdl.handle.net/10779/uow.27730308.v1

Lal M, Kalra DB. Lantana poisoning in domesticated animals. Indian Vet J. 1960;37:263–9.

Lone PA, Kothandaraman S, Dar JA, Hakeem KR, Khan ML. Invasive shrub (*Lantana camara* L.) alters the tree diversity and ecosystem-level carbon pools in tropical forests of Central India. Front For Glob Change. 2025;8:1412130. https://doi.org/10.3389/ffgc.2025.1412130.

Mack RN, Simberloff D, Lonsdale WM, Evans H, Clout M, Bazzaz FA. Biotic invasions: causes, epidemiology, global consequences, and control. Ecol Appl. 2000;10(3):689–710.

Mungi NA, Qureshi Q, Jhala YV. Expanding niche and degrading forests: key to the successful global invasion of *Lantana camara* (sensu lato). Glob Ecol Conserv. 2020;23:e01080.

Nawaz A, Ayub MA, Nadeem F, Al-Sabahi JN. Lantana (*Lantana camara*): a medicinal plant having high therapeutic potentials—a comprehensive review. Int J Chem Biochem Sci. 2016;10:52–9.

Negi GCS, Sharma S, Vishvakarma SC, Samant SS, Maikhuri RK, Prasad RC, Palni LMS. Ecology and use of *Lantana camara* in India. Bot Rev. 2019;85:109–30. https://doi.org/10.1007/s12229-019-09209-8.

Ntalo M, Ravhuhali KE, Moyo B, Hawu O, Msiza NH. *Lantana camara*: poisonous species and a potential browse species for goats in southern Africa – a review. Sustainability. 2022;14(2):751. https://doi.org/10.3390/su14020751.

Patel J, Qureshi MS, Kumar GS, Kumar D, Kumar KA. Phytochemicals and pharmacological activities of *Lantana camara* Linn. Res J Pharmacol Pharmacodyn. 2010;2(6):418–22. https://rjppd.org/AbstractView.aspx?PID=2010-2-6-27

Raphela TD, Duffy K. The impact of *Lantana camara* on invertebrates and plant species of the Groenkloof Nature Reserve, South Africa. Zool Stud. 2022;61:e33. https://doi.org/10.6620/ZS.2022.61-33.

Shackleton RT, Witt AB, Aool W, Pratt CF. Distribution of the invasive alien weed, *Lantana camara*, and its ecological and livelihood impacts in eastern Africa. Afr J Range Forage Sci. 2017;34(1):1–11. https://doi.org/10.2989/10220119.2017.1301551.

Sharma OP, Makkar HPS. *Lantana*-the foremost livestock killer in Kangra district of Himachal Pradesh. Livest Advis. 1981;6:29–31.

Sharma GP, Raghubanshi A. Effect of *Lantana camara* L. cover on local depletion of tree population in the Vindhyan tropical dry deciduous forest of India. Appl Ecol Environ Res. 2007;5:109–21.

Sharma O, Makkar HP, Dawra RK, Negi SS. A review of the toxicity of *Lantana camara* (Linn) in animals. Clin Toxicol. 1981;18(9):1077–94. https://doi.org/10.3109/15563658108990337.

Sharma OP, Sharma S, Pattabhi V, Mahato SB, Sharma PD. A review of the hepatotoxic plant *Lantana camara*. Crit Rev Toxicol. 2007;37(4):313–52. https://doi.org/10.1080/10408440601177863.

Sundaram B, Krishnan S, Hiremath A, Joseph G. Ecology and impacts of the invasive species, *Lantana camara*, in a social-ecological system in South India: perspectives from local knowledge. Hum Ecol. 2012;40:931–42. https://doi.org/10.1007/s10745-012-9532-1.

Tadele D. Allelopathic effects of Lantana (*Lantana camara* L) leaf extracts on germination and early growth of three agricultural crops in Ethiopia. Momona Ethiop J Sci. 2014;6:111–9.

William GL. Introduced plants, negative effects of *Lantana camara* (Verbanaceae). In: Levin SA, editor. Reference module in life sciences, encyclopaedia of biodiversity. 2nd ed. Academic Press; 2001. p. 345–56. https://doi.org/10.1016/B978-0-12-384719-5.00076-9. ISBN 9780123847201.

Wolfson SL, Solomons TWG. Poisoning by fruit of *Lantana camara*. An acute syndrome observed in children following ingestion of the green fruit. Am J Dis Child. 1964;107:173–6.

Chapter 6
Lantana camara: A Blessing and a Curse for Agriculture

Abstract An incredible adaptability, high fecundity, and rapid dispersal mechanisms have enabled *Lantana camara* to colonize a myriad of habitats and ecosystems and then extend into agricultural land. This chapter focuses on the good and bad aspects of *Lantana camara* in agroecosystems. In this chapter, its impacts and implications in agricultural activity; considering both the adverse impacts and the benefits (which are less commonly noted but still limited) of *Lantana camara* in agriculture will be discussed. The literature associated with ecological impacts and the allelopathic effects of *Lantana camara* and any known or potential effects on livestock will be examined. Lastly, the current management strategies implemented by agricultural managers to manage *Lantana camara* in agricultural systems will be reviewed.

6.1 Introduction: The Global Spread of an Ornamental Invader

Lantana camara or wild sage has been traveled globally for use as an ornamental plant due to its colorful and long-lasting flowers. It becomes an invasive species because of its rapid growth, enormous reproductive capacity (several thousands of viable seeds per year), and its ability to engage in both seed and vegetative propagation (rootstock, layering) (Negi et al. 2019). Its global movement from an ornamental garden plant to an agricultural pest has been addressed in previous chapters, and what makes it so invasive, consider its broad tolerance to ecological factors that allows for establishment in a great variety of climates and soils, yet most especially in an open environmental condition commonly experienced in agricultural settings (Skendžić et al. 2021).

Lantana camara is one of the most emerging invasive species in the tropical and subtropical parts of the world, especially parts of Asia, Africa, and Australia, as well as many locales of the Pacific Islands. Introduced initially for ornamental purposes, it has continued to encroach invasively, affecting the natural ecological setup of ecosystems and agricultural landscapes (Negi et al. 2019). *Lantana camara*, though

M. A. Dervash et al., *Lantana Camara*, SpringerBriefs in Plant Science,
https://doi.org/10.1007/978-3-032-03837-1_6

invasive in nature, contains numerous bioactive compounds used in agriculture for purposes of pest management and soil rehabilitation (Kumar et al. 2024).

This chapter offers a thorough review of *Lantana camara* within the scope of agricultural sciences. The ecological and economic impacts, management procedures, and possible utilizations of this plant within agricultural systems are discussed. By analyzing the interaction within agricultural ecosystems, the pros and cons of the plant are presented, with its detrimental effects, if any, and, more importantly, with its potential benefits within agricultural systems (Mhlongo et al. 2024).

6.2 The Curse: Detrimental Impacts on Agricultural Systems

Lantana camara has negative effects on agriculture, which are numerous and widely reported (Hamad et al. 2022). The various negative effects that make *Lantana camara* a curse for the agricultural industry are portrayed in Fig. 6.1.

This section will examine the key impacts it poses on agricultural productivity and sustainability:

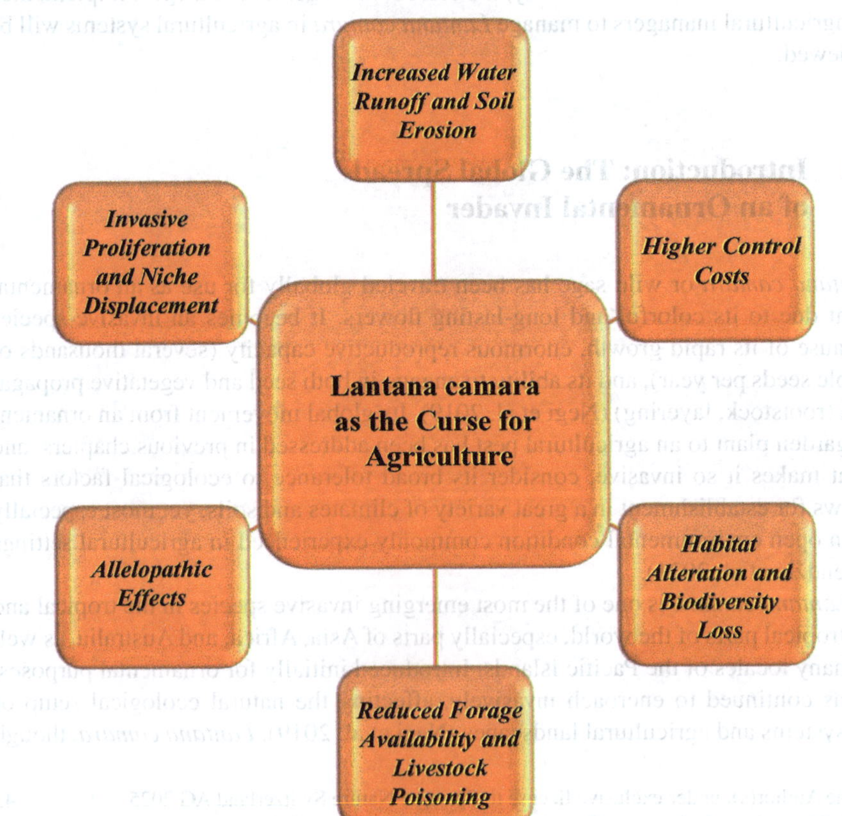

Fig. 6.1 Detrimental impacts of *Lantana camara* on agricultural systems

6.2.1   Invasive Proliferation and Niche Displacement

Lantana camara is recognized as an invasive and a bane to agricultural sectors. It grows rapidly with a high reproductive rate and adapts easily to multiple types of environments. These characteristics make it difficult for native plants and agricultural crops to thrive. *Lantana camara* can impact crop yield from over-competing for sunlight, water, and nutrients (Hamad et al. 2022). *Lantana camara* also takes space and resources in agricultural fields, pastures, orchards, and rangelands, hence causing huge economical losses for cattle-rearing industry owners. *Lantana camara* grows so well that it takes advantage of disturbed areas like roadsides, farm edges, barren plots, and wastelands. *Lantana camara* can grow and thrive in almost any type of soil, from dry to waterlogged soils. *Lantana camara* poses a great threat to crop production and biodiversity.

6.2.2   Allelopathic Effects

The other main reason for *Lantana camara*'s deleterious inhibiting agriculture is allelopathy. Allelopathic plant species release chemicals into the soil that prevent the germination and growth of plants surrounding them (Zheng et al. 2015). *Lantana camara* releases many different allelopathic chemicals, including flavonoids, terpenoids, and phenolics (e.g., umbelliferone, methylcoumarin, salicylic acid, and lantadenes), which can inhibit the growth of crops like maize, beans, and wheat negatively affecting agricultural productivity and yield by taking shots at crop's vigour; therefore, in areas where lantana is common, such allelopathic interference is very detrimental to farming (Ahmed et al. 2007).

6.2.3   Reduced Forage Availability and Livestock Poisoning

Lantana camara invades grazing fields and pastures, replacing valuable fodder plants. Moreover, *Lantana camara* has toxic pentacyclic triterpenoids (also known as lantadenes) in all parts of the plant, especially the leaves and immature berries. Ingestion by livestock types, like cattle, sheep, goats, horses, etc., leads to hepatotoxicity in animals, which is also manifested by photosensitization, jaundice, irritation of the skin, and, mostly, death (Sharma et al. 2007). This consequently leads to great economic losses to the farmers.

6.2.4   Habitat Alteration and Biodiversity Loss

In addition to direct competition, *Lantana camara* can also alter nutrient cycling and soil fertility, thereby rendering the environment congenial to its own establishment yet unfavorable to agricultural ecosystems in general (Panda et al. 2018). Further, thick stands may harbor vermin, snakes, and rodents, posing a threat to the farmer and his cattle.

6.2.5   Higher Control Costs

The widespread coverage of *Lantana camara* means that any control measures are expensive and labor-intensive, from hand-pulling, through mechanical slashing, grubbing, ploughing, and finally with chemical herbicides (Day et al. 2003). Even with all these efforts, full removal is difficult and, therefore, a burden that community members will continue to carry.

6.2.6   Increased Water Runoff and Soil Erosion

Some literature suggests that *Lantana camara* can mitigate soil compaction and erosion, while in dense thickets, it can increase surface water runoff and cause soil erosion, particularly where grass cover is poor (Rahman 2023).

6.3   The Blessing (or Glimmer of Hope): Potential Uses and Indirect Benefits

Lantana camara is largely a curse, but not without prospects, especially if its biomass is responsibly used. *Lantana camara* is also a glimmer of hope for the agricultural industry on account of potential uses and indirect benefits (Fig. 6.2).

This section will address the supposed "blessings" or benefits, which generally stem from attempts to control the weed.

6.3.1   Pollinator Attraction

The bright flowers attract many pollinators such as bees and butterflies, which may be beneficial for some agricultural crops in the vicinity that rely on insect pollination (Negi et al. 2019).

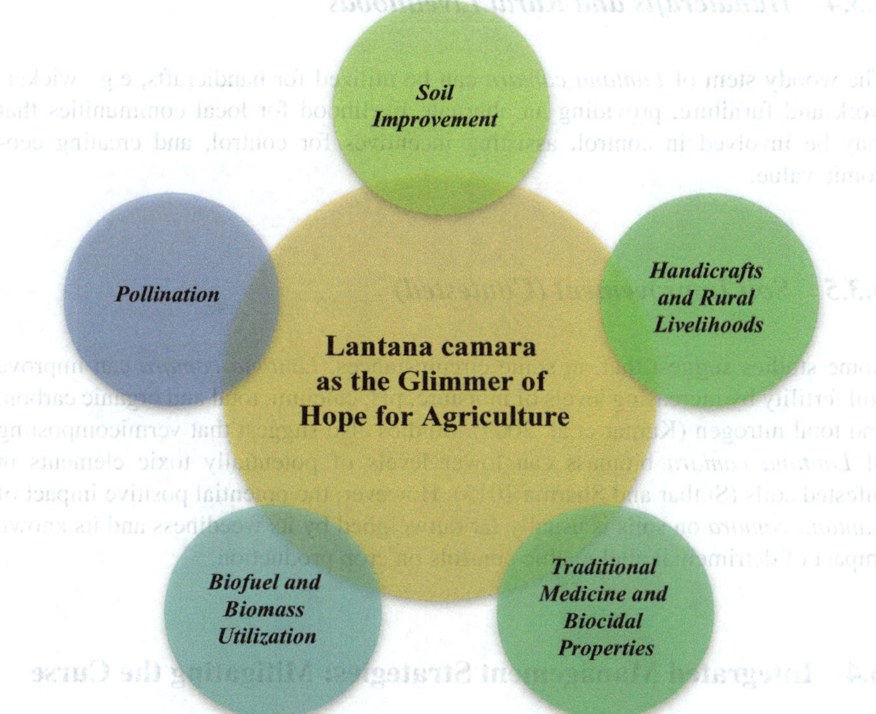

Fig. 6.2 *Lantana camara* as the glimmer of hope for agriculture

6.3.2   *Biofuel and Biomass Utilization*

The biomass of the plant can be used as fuel wood, specifically for cooking and heating in rural communities that lack alternative, energy-consuming methods (Singh et al. 2023). Research has investigated the plant for biogas generation, paper pulp (the plant stems possess furfuroid, lignin, cellulose), and possibly roofing materials.

6.3.3   *Traditional Medicine and Biocidal Properties*

Lantana camara may have historically been used in traditional medicine with a number of attributes as an antimicrobial, fungicide, insecticide, nematicide, anti-inflammatory, and antioxidant (Naz and Bano 2013). The extracts for the plant have shown activity against bacteria, fungi, and parasites. Furthermore, some research has suggested the presence of anticancer properties within *Lantana camara* plant extracts. As always, caution should be used in this area due to the inherent toxicity of the plant.

6.3.4 Handicrafts and Rural Livelihoods

The woody stem of *Lantana camara* can be utilized for handicrafts, e.g., wickerwork and furniture, providing an alternate livelihood for local communities that may be involved in control, assisting incentives for control, and creating economic value.

6.3.5 Soil Improvement (Contested)

Some studies suggest that, in some circumstances, *Lantana camara* can improve soil fertility by increasing levels of moisture, pH, calcium, total and organic carbon, and total nitrogen (Kumar et al. 2009). Studies also suggest that vermicomposting of *Lantana camara* biomass can lower levels of potentially toxic elements in infested soils (Suthar and Sharma 2013). However, the potential positive impact of *Lantana camara* on soils is usually far outweighed by its weediness and its known impact of detrimental allelopathic controls on crop production.

6.4 Integrated Management Strategies: Mitigating the Curse

Lantana camara is an opportunistic species that will require an integrated strategy for effective management, as viewed from a long-term perspective and efficiency of effort. In this relation, some of the current and emerging methods of control are mentioned, with an emphasis on methods that are specifically relevant to agricultural contexts (Nanjappa et al. 2005):

6.4.1 Mechanical/Manual Control

Hand-pulling seedlings, cutting and uprooting larger plants, and using machinery like slashers, bulldozers, and ploughs are common methods. However, these are often labor-intensive and require thorough root removal to prevent regrowth.

6.4.2 Chemical Control

Herbicides (including 2,4-D + picloram, triclopyr, and metsulfuron) are sold quite widely and are typically applied as foliar herbicides and as either a basal bark treatment or cut-stump. Timing the application and manner of applying the herbicide is essential for effectiveness and limiting environmental impacts. Herbicides do provide a means of control, but concern over environmental impacts, effects on nontarget species, and that *Lantana camara* may develop herbicide-resistance still remains. With this in mind, the use of herbicides can be deliberately incorporated within an integrated pest management (IPM) approach.

6.4.3 Biological Control

There have been a number of different biological control agents (insects, fungi) that have been examined in depending locations, but little success has been achieved in reducing, to the extent and duration required for population control, the *Lantana camara*, given the adaptability of this plant to its environment and genetic diversity.

6.4.4 Ecological Restoration and Prevention

Long-term success is rarely achieved when managing *Lantana camara* in agricultural landscapes. Effective management often considers the ecological restoration of weed-free areas with native or suitable pasture species. Restoration can help minimize reinvasion of an area. Also, preventing access to *Lantana camara* as an ornamental plant through prohibition and monitoring weed-free areas are key preventative components.

6.4.5 Community-Based Management

Supporting local communities for awareness campaigns, creating livelihood possibilities through *Lantana camara,* and developing comanagement plans are essential for sustainable management.

6.5 Conclusion

Lantana camara is an incredibly serious threat to agricultural ecosystems globally, affecting crop production, livestock welfare, and biodiversity. It possesses a serious invasive status that outweighs its limited advantageous uses. Although there are some minor positive opportunities to provide positive incentives from it through biomass use, it offers little or no solutions and cannot provide solutions in the absence of a credible-based, integrated management plan. The closing section of the chapter reiterates that additional research toward improved sustainability in cost-efficient control methods is paramount toward effective integrated control, and stresses that biocontrol needs to be extended to include eradication through ecological restoration from which the community is involved. Essentially, while *Lantana camara* is uncontested as an invasive species in agriculture, it will continue to be seen predominantly as a curse requiring persistent vigilance and management to counteract its ongoing deleterious effects.

References

Ahmed R, Uddin MB, Khan MA, Mukul SA, Hossain MK. Allelopathic effects of *Lantana camara* on germination and growth behavior of some agricultural crops in Bangladesh. J For Res. 2007;18:301–4.

Day MD, Wiley CJ, Playford J, Zalucki MP. *Lantana* current management status and future prospects. Canberra: ACIAR Monograph; 2003.

Hamad AA, Kashaigili JJ, Eckert S, Eschen R, Schaffner U, Mbwambo JR. Impact of invasive *Lantana camara* on maize and cassava growth in East Usambara, Tanzania. Plant Environ Interact. 2022;3(5):193–202. https://doi.org/10.1002/pei3.10090.

Kumar P, Pant M, Negi GCS. Soil physico-chemical properties and crop yield improvement following *Lantana* mulching and reduced tillage in rainfed croplands in the Indian Himalayan mountains. J Sustain Agric. 2009;33:636–57.

Kumar R, Guleria N, Deeksha MG, Kumari N, Kumar R, Jha AK, Parmar N, Ganguly P, de Aguiar Andrade EH, Ferreira OO, de Oliveira MS. From an invasive weed to an insecticidal agent: exploring the potential of *Lantana camara* in insect management strategies-a review. Int J Mol Sci. 2024;25(23):12788. https://doi.org/10.3390/ijms252312788.

Mhlongo ES, Ruwanza S, Dalu T. Perceptions, knowledge, and invasion extent of *Lantana camara* on household yards in rural communities in Limpopo province, South Africa. Soc Nat Resour. 2024;37(8):1218–39. https://doi.org/10.1080/08941920.2024.2338773.

Nanjappa HV, Saravanane P, Ramachandrappa BK. Biology and management of Lantana camara L.–a review. Agric Rev. 2005;26(4):272–80.

Naz R, Bano A. Phytochemical screening, antioxidants and antimicrobial potential of *Lantana camara* in different solvents. Asian Pacific J Trop Dis. 2013;3(6):480–6. https://doi.org/10.1016/S2222-1808(13)60104-8.

Negi GCS, Sharma S, Vishvakarma SC, Samant SS, Maikhuri RK, Prasad RC, Palni LMS. Ecology and use of *Lantana camara* in India. Bot Rev. 2019;85:109–30. https://doi.org/10.1007/s12229-019-09209-8.

Panda T, Mishra N, Pradhan BK, Mohanty RB. Expansive alien flora of Odisha, India. JAEID. 2018;112(1):43–64.

Rahman MM. Assessing the role of invasive Lantana (*Lantana camara*) in disrupting the native biodiversity of Bangladesh (April 6, 2023). Available at SSRN: https://ssrn.com/abstract=4411276 or https://doi.org/10.2139/ssrn.4411276 (2023)

Sharma OP, Sharma S, Pattabhi V, Mahato SB, Sharma PD. A review of the hepatotoxic plant *Lantana camara*. Crit Rev Toxicol. 2007;37(4):313–52. https://doi.org/10.1080/10408440601177863.

Singh AK, Pal P, Rathore SS, Sahoo UK, Sarangi PK, Prus P, Dziekański P. Sustainable utilization of biowaste resources for biogas production to meet rural bioenergy requirements. Energies. 2023;16(14):5409. https://doi.org/10.3390/en16145409.

Skendžić S, Zovko M, Živković IP, Lešić V, Lemić D. The impact of climate change on agricultural insect pests. Insects. 2021;12(5):440. https://doi.org/10.3390/insects12050440.

Suthar S, Sharma P. Vermicomposting of toxic weed-*Lantana camara* biomass: chemical and microbial properties changes and assessment of toxicity of end product using seed bioassay. Ecotoxicol Environ Saf. 2013;95:179–87. https://doi.org/10.1016/j.ecoenv.2013.05.034.

Zheng YL, Feng YL, Zhang LK, Callaway RM, Valiente-Banuet A, Luo DQ. Integrating novel chemical weapons and evolutionarily increased competitive ability in success of a tropical invader. New Phytol. 2015;205:1350–9.

Chapter 7
Traditional Medicinal Uses and Scientific Validation of *Lantana camara*

Abstract *Lantana camara*, a common invasive weed, has, for reasons that have still to be conclusively determined, developed a plethora of traditional medicinal uses. This chapter has highlighted the multifaceted duality of this plant. *Lantana camara* is presently used as folk medicine for various ailments, including inflammation, pain, infections, fever, and respiratory issues, and it has a lofty place in traditional folk medicine as an antimalarial agent, where indigenous people rely on the use of *Lantana camara* with confidence. Phytochemical studies have identified many diverse compounds, including a large number of triterpenoids, especially lantadenes, flavonoids, and essential oils that have also been isolated. Studies in the scientific literature have shown good support for its anti-inflammatory, analgesic, antimicrobial, antimalarial, and antioxidant properties. That said, the intrinsic toxicity of *Lantana camara* is its primary deterrent in studying its medicinal properties, associated with toxicity mainly due to lantadenes. A significant focus for future research should involve following similar steps to other plant-based medicines, including standardization and safety of what extracts are tested, isolation of bioactive components known to be safe, and clinical testing in developing antisepsis, antimicrobial, anti-inflammatory, and analgesic properties. Ultimately, with an appropriate understanding of *Lantana camara*, invasive species may be turned into an opportunity to develop a source of new therapies, bridging traditional ways of knowing with modern pharmacology.

7.1   Introduction

Lantana camara, a colorful and abundant shrub of the family Verbenaceae, is a unique juxtaposition. It is considered one of the world's most invasive and troublesome weeds, as its combative growth and allelopathic characteristics are serious threats to biodiversity and agriculture in the tropical and subtropical zone of the planet (Negi et al. 2019). Surprisingly and unfortunately, the obvious environmental negative factors often overshadow its historical roles in culturally based traditional medicine, which have been a part of the pharmacopeia of various cultures around the globe for some time. This chapter is intended to clarify the contradictory character of *Lantana camara* with respect to folk medicine structure and to consider the evidence that validates or questions the traditional knowledge surrounding *Lantana camara*.

   Lantana camara, which is originally from the American tropics, is now considered to have a global distribution that can tolerate a mix of climates and soil types. It's often multicolored flowers, which cluster together, will of course be easy to identify, as will its aromatic plant parts. While it has been proven toxic to livestock, humans have recognized the potential advantages of utilizing different plant parts for specific treatments for centuries (Kumar and Singh 2020). To accurately assess the potential true effects and dangers, it is very important to understand the historical context involved and conduct careful scientific evaluation.

7.2 Traditional Medicinal Uses of *Lantana camara*

Lantana camara has been used in traditional medicine in many ways, in many places across many cultures, highlighting the plant's global distribution and incorporation into local cultures. From the leaves, flowers, roots, and fruit, many traditional practitioners have used many parts of the plant to treat a diverse array of medical conditions and often in different ways: as a decoction, dried for an infusion, as a poultice, and as a paste (Kumar et al. 2022). In this section, some of the more common traditional uses of *Lantana camara* are highlighted.

7.2.1 Anti-inflammatory and Analgesic Uses

The most common traditional use of *Lantana camara* is to treat inflammation and relieve pain. In many areas of Africa, Asia, and Latin America, fresh leaves or dried leaves were crushed and applied to the skin in a poultice for swelling, sprains, bruises, and rheumatism. Similarly, decoction of leaves was ingested for joint or muscle pain and swelling. According to the traditional beliefs, *Lantana camara* has properties believed to reduce inflammation to relieve pain symptoms and aid symptoms such as swelling (Khairan et al. 2024).

7.2.2 Antimicrobial and Antiseptic Uses

Lantana camara is known to have a strong traditional reputation as an antimicrobial agent. In folk medicine, the leaf extracts are often used to clean and disinfect wounds, cuts, and skin infections, as well as to treat boils, ulcers, and parasitic skin conditions such as scabies (Nayak et al. 2009). Some systems use the leaf decoctions as gargles for sore throats and mouth infections and as washes for eye infections. These diverse applications indicate a widely held belief concerning the plant's ability to combat microbial pathogens and support healing (Barreto et al. 2010).

7.2.3   Antipyretic Action

Fevers, as a common symptom of different ailments, have also been treated in a traditional way using *Lantana camara*. In many places, especially in communities in India and Southeast Asia, an infusion or decoction of leaves is given orally to treat fever. The leaves are believed to induce sweating, which would assist in lowering body temperature (Nawaz et al. 2016).

7.2.4   Respiratory Ailments

Traditional healing practices have also reported the use of *Lantana camara* for respiratory conditions. Leaves and flower decoctions are typically used for symptoms of colds, coughs, bronchitis, and asthma. They are thought to have some expectorant and bronchodilatory effects, and they can open up airways and assist with breathing (Nawaz et al. 2016).

7.2.5   Gastrointestinal Disorders

Gastroenterological applications are much less common due to concerns about toxicity, but some forms of traditional applications do exist within the healing traditions. In some regions, dilute decoctions are used to treat diarrhea, dysentery, and stomach pains, but this application is often accompanied by the caution that there can be gastrointestinal irritation when using larger doses of the plant (Sagar et al. 2005).

7.2.6   Skin Conditions and Dermatological Uses

Lantana camara has been used in traditional practices as a skin treatment beyond that of wound healing and care. Traditional uses include treating problems, such as eczema, ringworm, and other fungal manifestations. The leaves are commonly macerated and applied topically to the afflicted skin, or infusions of the plant have been made in oils to use topically (Negi et al. 2019).

7.2.7 *Antimalarial Properties*

Lantana camara has been used in some malaria-endemic regions as a traditional antimalarial agent. Infusions or decoctions of the leaves have been ingested to relieve fever and other symptoms of malaria. This particular use, as a traditional antimalarial, has gained a considerable amount of scientific attention (Kacholi 2024).

7.3 Scientific Validation of *Lantana camara*: Bridging Tradition and Modern Science

The introduction of modern scientific methods has enabled researchers to evaluate the pharmacological foundation of the traditional medical claims. Many studies have been performed on the medicinal plant, *Lantana camara*, to identify the chemical components and evaluate biological activities (Sharma et al. 2025). This section will present scientific evidence that supports or challenges the traditional uses highlighted above.

7.3.1 *Anti-inflammatory and Analgesic Activity: Scientific Confirmation*

Multiple studies in vitro and in vivo have demonstrated a confirmation of the historical use of *Lantana camara* for its anti-inflammatory and analgesic effects.

- *In vitro studies: Lantana camara* extracts (ethanolic and aqueous) have demonstrated the inhibition of pro-inflammatory mediators (in liver and macrophage cell lines), including prostaglandins, leukotrienes, and nitric oxide. These studies indicate that there might be a mechanism of action involving the modulation of inflammatory pathways (Wu et al. 2020; Khairan et al. 2024).
- *In vivo studies: Lantana camara* extracts produced a pain reduction (in the form of reduced swelling and inflammation) in several animal models of inflammation (carrageenan-induced paw edema and cotton pellet granuloma). The analgesia is of two main types: peripheral and that which may result from a central locus of drug action (e.g., inhibition of pain mediators and perhaps an opioid pathway).
- *Key compounds:* There are flavonoids and specific triterpenoids that are the contributors to these anti-inflammatory and analgesic effects that counteract inflammatory pathways caused by pro-inflammatory compounds.

7.3.2 Antimicrobial Activity: A Strong Scientific Basis

Scientific research backs up the traditional use of *Lantana camara* as an antimicrobial.

- *Broad-spectrum activity:* Extracts from whole *Lantana camara* (leaves, flowers, essential oil) have shown inhibitory efficacy against a wide variety of bacteria (Gram-positive and Gram-negative) and fungi (Sharma and Kumar 2009).
- *Bacterial targets:* In vitro studies have reported efficacy against many pathogenic bacteria, including *Staphylococcus aureus*, *Escherichia coli*, and *Proteus mirabilis* (Sharma and Kumar 2009).
- *Fungal targets:* Antifungal activity has been reported with respect to *Candida albicans* and *Trichophyton mentagrophytes* (Sharma and Kumar 2009).
- *Mechanism of action:* Antimicrobial activity is associated with flavonoids, triterpenoids, and compounds found in the essential oil (e.g., β-caryophyllene, humulene), which result in disruption of bacterial cell membranes, inhibition of enzyme activity, and interference with microbial growth, growth, and/or replication (Sharma and Kumar 2009).

7.3.3 Antipyretic Activity: Limited but Promising Evidence

Although it has been less rigorously studied than its anti-inflammatory effects, some researchers have validated its traditional antipyretic use. Different animal studies have identified certain extracts as effective in reducing elevated body temperatures induced by pyrogens. The mechanism is not yet clear, though one possibility is that *Lantana camara* might inhibit prostaglandin synthesis in the hypothalamus (Insel 1990).

7.3.4 Antimalarial Activity: A Focus of Research

The traditional use of *Lantana camara* in areas of endemic malaria has resulted in some interest in discovering its antimalarial properties.

- *In vitro studies:* There have been studies that have reported that extracts of *Lantana camara* inhibited some *Plasmodium falciparum* strains (the parasite that causes most cases of human malaria) in in vitro assays (Kacholi 2024; Sonter et al. 2024).
- *In vivo studies:* There are some malaria animal models that experienced a drop in parasitemia and an increase in survival rate after *Lantana camara* extract (Dua et al. 2010).

- *Active compounds:* No strong antimalarial compound(s) have been found to possess commercial potential comparable to artemisinin or quinine, but there may be several compounds (flavonoids, triterpenoids, etc.) with synergistic activity (Czechowski et al. 2019). Research continues to identify and isolate those antimalarial compounds and mechanisms of action, etc.

7.3.5 Antioxidant Activity: A General Protective Mechanism

Various studies have reinforced the strong antioxidant activity noted in the extracts of *Lantana camara,* especially in extracts with high flavonoid and phenolic compounds (Kumar et al. 2014).

- *Free radical scavenging:* Extracts were confirmed to effectively scavenge a variety of free radicals, including DPPH, ABTS, and superoxide radicals (Mansoori et al. 2020).
- *Reducing power:* The extracts also exhibited good reducing power, indicating their ability to donate electrons or neutralize reactive oxygen species (Mansoori et al. 2020).
- *Implications:* This antioxidant activity adds to the overall protective effects of the plant on oxidative stress and ultimately has implications in many chronic diseases and inflammation. It may also relate to its wound healing activity, as well as its anti-inflammatory actions (Khairan et al. 2024; Mansoori et al. 2020).

7.3.6 Supplementary Documented Pharmacological Activities

In addition to the confirmation of traditional uses, scientific investigations have found additional pharmacological activities of *Lantana camara* that may be interesting:

- *Anticancer activity:* Recent preliminary in vitro studies have also found that extracts of *Lantana camara* and isolated compounds have cytotoxicity against different cancer cell lines. This area is a budding area for exploration—much more research would be needed to establish this relationship (Babar et al. 2019; El Hajj et al. 2024).
- *Insecticidal and larvicidal activity:* The essential oil and some extracts of *Lantana camara* have produced strong insecticidal and larvicidal effects against mosquitoes and agricultural pests. Again, this confirms a traditional use of it, providing an insect repellent effect (Rajashekar et al. 2014).
- *Antidiabetic activity:* A handful of studies have suggested that *Lantana camara* may also have a role to play in regulating blood glucose concentrations; these studies allude to some effect on carbohydrate metabolism (Kapali and Sharma 2021; El Hajj et al. 2024).

7.3.7 Toxicity and Safety Concerns: A Critical Examination

Although the scientific validation of a number of traditional uses is encouraging, it must also be acknowledged that *Lantana camara* has a documented toxicity, especially in livestock (Sharma et al. 2007).

- *Lantadenes:* The primary toxic compounds are pentacyclic triterpenoids, which produce cholestasis, liver damage, and photosensitivity in grazing animals, and account for large economic loss (Pass et al. 1979).
- *Human toxicity:* Cases of *Lantana camara* poisoning in humans are less common but have been reported. Poisoning mostly involves ingestion of unripe fruits, which contain a large concentration of toxic compounds, and associated symptoms are reported as nausea, vomiting, diarrhoea, abdominal pain, and, in severe cases, neurological damage (Carstairs et al. 2010; Sharma et al. 2007).
- *Dose-dependency:* The range of *Lantana camara* in humans is seen to a limited extent at best, and the line between harmful and helpful is generally conditioned by the exact extract, dose, and route of administration (Carstairs et al. 2010; Sharma et al. 2007).
- *Variability:* The concentration of *Lantana camara* induced toxicity may vary widely based on the genotype of the selected plant, growth phase, environmental conditions, and the maturity.
- *Traditional knowledge and detoxification:* It is crucial to consider that traditional healers often take advantage of specific preparations (e.g., decoction, method of drying, etc.) that alter the concentration of toxic compounds or lessen their effects; certainly this begs further scientific validation.
- *Research implications:* Future research should focus on isolating the beneficial compounds alone, free from toxic lantadenes, or preparing standardized extracts with defined safety and therapeutic indices.

7.4 Conclusion

Lantana camara has earned the notoriety of being an invasive weed. There are ample opportunities for this plant to become sources of beneficial molecules. Traditional knowledge has afforded a crucial map for researchers who are beginning to catalogue the ways in which *Lantana camara* is used to treat inflammatory responses, infections, and fevers. The idea that Lantana camara would be efficacious against malaria, and possibly as an antioxidant, has been supported by just a handful of studies. The plant has known toxicity and therefore should use caution and good science when using it for medicine. Careful work in phytochemistry research, pharmacology, and toxicology can help uncover the therapeutic benefits of this unique and contradictory plant, which could help make a global problem into a source of new and useful therapies. The future of *Lantana camara* in medicine lies in rigorous and scientific study to help determine the benefits of using the plant safely as remedies for human health.

References

Babar VB, Khapale PR, Nagarale S. Preliminary investigation and in-vitro anticancer activity of *Lantana camara* L. (Verbenaceae). J Pharm Phytochem. 2019;8:2524–7.

Barreto F, Sousa E, Campos A, Costa J, Rodrigues F. Antibacterial activity of *Lantana camara* Linn and Lantana montevidensis brig extracts from Cariri-Ceará, Brazil. J Young Pharm. 2010;2(1):42–4. https://doi.org/10.4103/0975-1483.62211.

Carstairs SD, Luk JY, Tomaszewski CA, Cantrell FL. Ingestion of *Lantana camara* is not associated with significant effects in children. Pediatrics. 2010;126(6):1585–8. https://doi.org/10.1542/peds.2010-1669.

Czechowski T, Rinaldi MA, Famodimu MT, Van Veelen M, Larson TR, Winzer T, Rathbone DA, Harvey D, Horrocks P, Graham IA. Flavonoid versus artemisinin anti-malarial activity in artemisia annua whole-leaf extracts. Front Plant Sci. 2019;10:984. https://doi.org/10.3389/fpls.2019.00984.

Dua VK, Pandey AC, Dash AP. Adulticidal activity of essential oil of *Lantana camara* leaves against mosquitoes. Indian J Med Res. 2010;131(3):434–9.

El Hajj J, Karam L, Jaber A, Cheble E, Akoury E, Kobeissy PH, Ibrahim JN, Yassin A. Evaluation of antiproliferative potentials associated with the volatile compounds of *Lantana camara* flowers: selective in vitro activity. Molecules. 2024;29(22):5431. https://doi.org/10.3390/molecules29225431.

Insel PA. Analgesic-antipyretic and anti-inflammatory agents and drugs employed in the treatment of gout. In: Hardman JG, Limbird LE, editors. The pharmacological basis of therapeutic, vol. 1990. McGraw Hill; 1990. p. 617–58.

Kacholi DS. A comprehensive review of antimalarial medicinal plants used by Tanzanians. Pharm Biol. 2024;62(1):133–52. https://doi.org/10.1080/13880209.2024.2305453.

Kapali J, Sharma KR. Estimation of phytochemicals, antioxidant, antidiabetic and brine shrimp lethality activities of some medicinal plants growing in Nepal. J Med Plants. 2021;20(80):102–16.

Khairan K, Maulydia NB, Faddillah V, Tallei TE, Fauzi FM, Idroes R. Uncovering anti-inflammatory potential of *Lantana camara* Linn: network pharmacology and in vitro studies. Narra J. 2024;4(2):894. https://doi.org/10.52225/narra.v4i2.894.

Kumar RP, Singh JS. Invasive alien plant species: their impact on environment, ecosystem services and human health. Ecol Indic. 2020;111:106020. https://doi.org/10.1016/j.ecolind.2019.106020.

Kumar S, Sandhir R, Ojha S. Evaluation of antioxidant activity and total phenol in different varieties of *Lantana camara* leaves. BMC Res Notes. 2014;7:560.

Kumar A, Singh S, Kewat A, Chand H. *Lantana camara*: an obnoxious weed with therapeutic potential promoting sustainable rural livelihood and drug discovery. In: Biodiversity in the service of mankind. Walnut Publication; 2022.

Mansoori A, Singh N, Dubey SK, Thakur TK, Alkan N, Das SN, Kumar A. Phytochemical characterization and assessment of crude extracts from *Lantana camara* L. for antioxidant and antimicrobial activity. Front Agron. 2020;2:582268. https://doi.org/10.3389/fagro.2020.582268.

Nawaz A, Ayub MA, Nadeem F, Al-Sabahi J. Lantana (*Lantana camara*): a medicinal plant having high therapeutic potentials-a comprehensive review. Int J Chem Biochem Sci. 2016;10:52–9.

Nayak BS, Raju SS, Eversley M, Ramsubhag A. Evaluation of wound healing activity of *Lantana camara* L. – a preclinical study. Phytother Res. 2009;23(2):241–5. https://doi.org/10.1002/ptr.2599.

Negi GCS, Sharma S, Vishvakarma SCR, Samant SS, Maikhuri RK, Prasad RC, Palni LMS. Ecology and use of *Lantana camara* in India. Bot Rev. 2019;85:109–30. https://doi.org/10.1007/s12229-019-09209-8.

Pass MA, Seawright AA, Lamberton JA, Heath TJ. Lantadene a toxicity in sheep. A model for cholestasis. Pathology. 1979;11(1):89–94. https://doi.org/10.3109/00313027909063543.

Rajashekar Y, Ravindra KV, Bakthavatsalam I. Leaves of *Lantana camara* Linn. (Verbenaceae) as a potential insecticide for the management of three species of stored grain insect pests. J Food Sci Technol. 2014;51(11):3494–9. https://doi.org/10.1007/s13197-012-0884-8.

Sagar L, Sehgal R, Ojha S. Evaluation of antimotility effect of *Lantana camara* L. var. acuelata constituents on neostigmine induced gastrointestinal transit in mice. BMC Complement Altern Med. 2005;5:1–6.

Sharma B, Kumar P. Bioefficacy of *Lantana camara* L. against some human pathogens. Indian J Pharm Sci. 2009;71(5):589–93. https://doi.org/10.4103/0250-474X.58177.

Sharma OP, Sharma S, Pattabhi V, Mahato SB, Sharma PD. A review of the hepatotoxic plant *Lantana camara*. Crit Rev Toxicol. 2007;37:313–52. https://doi.org/10.1080/10408440601177863.

Sharma B, Kumar A, Kumar R, Sunita K, Jena N, Kumar S. Evaluation of medicinal values and phytochemical analysis of *Lantana camara* flowers: identification of bioactive compounds. In: Plants and secondary metabolites, vol. 6. Ambika Prasad Research Foundation; 2025.

Sonter S, Dwivedi MK, Mishra S, Singh P, Kumar R, Park S, Jeon BH, Singh PK. In vitro larvicidal efficacy of *Lantana camara* essential oil and its nano-emulsion and enzyme inhibition kinetics against *Anopheles culicifacies*. Sci Rep. 2024;14(1):16325. https://doi.org/10.1038/s41598-024-67148-w.

Wu P, Song Z, Wang X, Li Y, Li Y, Cui J, Tuerhong M, Jin D, Abudukeremu M, Lee D, Xu J, Guo Y. Bioactive triterpenoids from *Lantana camara* showing anti-inflammatory activities in vitro and in vivo. Bioorg Chem. 2020;101:104004.

Chapter 8
Lantana camara: A Paradoxical Pharmacopoeia

Abstract This chapter reviews the uniqueness of *Lantana camara*, examining its extensive ethnobotanical uses for conditions ranging from inflammatory responses and infections to respiratory and gastrointestinal distress. The chapter reviews the vast phytochemistry, identifying important bioactive compounds, such as triterpenoids (the toxic lantadenes and the beneficial oleanolic acid and ursolic acid), flavonoids, alkaloids, saponins, and essential oils, that could have therapeutic effects. This chapter also weighs the important problem of inherent toxicity, including hepatotoxicity and photosensitization for livestock (and danger to humans) from the lantadenes. By extension, there is a need for scientific research to isolate beneficial compounds and decrease toxicity to produce safe, standardized medications from this complicated plant. There is a paradox here, which makes *Lantana camara* very powerful medicinally.

8.1 Introduction

Lantana camara L., belonging to the family Verbenaceae, represents a botanical contradiction. Universally viewed as one of the ten most invasive weeds in the world, *Lantana camara* represents a serious threat to the ecology and agriculture of tropical and subtropical areas. Its capacity for rapid growth, allelopathy, and generalist nature have established its formidable adversarial reputation for both conservationists and farmers alike. Underneath this ubiquity of *invasiveness*, however, exists a treasure trove of phytochemistry, a long use in traditional medicine, and an accumulating scientific literature suggesting therapeutic value (Kato-Noguchi and Kato 2025).

In this chapter, *Lantana camara* in all its facets, including its ethnobotanical history, phytochemistry, and scientific evidence in support of medicinal effects, is explored (Ramírez et al. 2025). At the same time, we will highlight the potential toxicity that should be regarded when taking advantage of this medication and

M. A. Dervash et al., *Lantana Camara*, SpringerBriefs in Plant Science,
https://doi.org/10.1007/978-3-032-03837-1_8

63

thereby create an all-encompassing understanding of *Lantana camara* as a true *medicinal warrior*, powerful and abundant, yet demanding respect for its formidable double-edged nature.

8.2 Ethnobotanical Heritage: A Legacy of Healing Across Cultures

Medicinal use of plants predates recorded history, and *Lantana camara* is no exception. Indigenous cultures and local practitioners around the world have long recognized and utilized the therapeutic benefits of this seemingly ubiquitous shrub. Its global distribution, part of which is due to its invasive qualities, has permitted integration into a number of ethnomedicinal systems, often for remarkably similar conditions (Kato-Noguchi and Kato 2025).

8.2.1 Anti-Inflammatory and Analgesic Uses

The most consistent traditional usage of *Lantana camara* focuses on its perceived ability to reduce pain and inflammation (El-Banna et al. 2022). Fresh or dried leaves were applied in Africa and Asia, as well as Latin America, as poultices or decocted, to treat arthritis or rheumatic conditions, muscle sprains, or general body aches. Crushed leaves were applied topically, with the belief that it would reduce swelling and immobilize pain (Wu et al. 2020).

8.2.2 Antimicrobial and Antiseptic Uses

The treatment of infections is another reoccurring idea. Historically, the leaves of *Lantana camara* were macerated and then applied directly to wounds, cuts, ulcers, and a number of skin diseases, including scabies, leprosy lesions, and eczema (Nayak et al. 2009). Presumably, the leaves possessed some antiseptic properties that were necessary to prevent infection when modern-day antibiotics were not available. The essential oil derived from *Lantana camara* has also been historically applied in this way for antimicrobial properties (Nayak et al. 2009).

8.2.3 Respiratory and Gastrointestinal Remedies

As respiratory complaints were the most common use of *Lantana camara*, traditional practitioners used decoctions of it to relieve coughs, colds, bronchitis, asthma, and, possibly, catarrhal infections (suggesting that it is likely an expectorant or a bronchodilator) (Jabeen et al. 2009). Traditional practitioners also used it to treat gastrointestinal complaints including stomach aches, dysentery, and to expel intestinal parasites (vermicides) using infusions of its leaves or roots (Tadesse et al. 2017). This last use is an indication of its potential efficacy as an anthelminthic agent (Sagar et al. 2005).

8.3 Phytochemical Canvas: Unveiling the Bioactive Compounds

The high number of pharmacological activities attributed to *Lantana camara* is due to the many possible phytochemical composites available in the plant. All parts of the plant are phytochemically rich, leaves, stems, roots, flowers, and berries, of which the type and abundance of phytochemicals will differ by part of the plant, origin, environment, and genotype.

8.3.1 Triterpenoids

This group of agents may be the most important, as this group holds the plant's therapeutic agents against breast cancer (Bishayee et al. 2011).

8.3.2 Lantadenes (Lantadene A, B, C, D, E)

These are pentacyclic triterpenoids and are to blame for the hepatotoxicity and photosensitization in livestock. Lantadene A and B and their isomers are relatively abundant and active. Therefore, knowing the appropriate presence and concentration for safety is important (Kumar et al. 2018).

8.3.3 Oleanolic Acid and Ursolic Acid

These are triterpenoids that are nearly ubiquitous and are well-known for their potent anti-inflammatory, antioxidant, anticancer, and hepatoprotective activity (Wu et al. 2020).

8.3.4   Flavonoids

Flavonoids, which are polyphenolic compounds, are known to possess strong antioxidant, anti-inflammatory, antiviral, and anticarcinogenic properties (Zahra et al. 2024). *Lantana camara* contains a variety of flavonoids that contribute to the overall therapeutic effects of *Lantana camara* as both a combatant of oxidative stress and inflammation (Kakade et al. 2025).

8.3.5   Alkaloids as a Diverse Range of Bioactivity

Alkaloids, a class of nitrogen-containing organic compounds, are typically less abundant than triterpenoids and, as a class, have a profound effect on physiological processes. The exact role of alkaloids in *Lantana camara*'s pharmacologic effects is currently being studied (Al-Snafi 2019).

8.3.6   Saponins as Multifunctional Glycosides

Saponins, a type of glycoside, are characterized as foaming agents and have been ascribed a number of biological activities, including anti-inflammatory, cholesterol-lowering, antimicrobial activities, and used in anticancer drugs (Elekofehinti et al. 2021). The presence of saponins suggests yet another mechanism by which *Lantana camara* may exert its multitude of therapeutic actions.

8.3.7   Steroids as Hormonal Precursors
*   and Bioactive Compounds*

Steroids found in plants, also known as phytosterols, potentially have a range of biological activities, such as anti-inflammatory activity and endocrine activity modulation (Shen et al. 2024).

8.3.8   Tannins as Astringents and Antioxidants

Tannin is a polyphenolic compound that has astringent properties, which can induce some localized healing and wound healing, and also the antimicrobial potential. Tannins also have a relevant antioxidant capacity (Lin et al. 2016).

8.3.9   *Essential Oils (Aromatic and Antimicrobial Constituents)*

The volatile essential oil isolated from *Lantana camara* is a mixture of monoterpenes and sesquiterpenes. In particular, the essential oil includes β-caryophyllene, spathulenol, γ-cadinene, trans-β-farnesene, p-cymene, and a range of alcohol constituents, including cis-3-hexen-1-ol, 1-octen-3-ol, caryophyllene oxide, and 1-hexanol (Khan et al. 2016). These essential oils represent the characteristic smell and range of key antimicrobial, insecticide, and anti-inflammatory properties of the plant. Principal chemical constituents of the essential oils can differ significantly depending on the chemotype, which also may affect the therapeutic application or safety of the essential oil (Khan et al. 2016).

8.3.10   *Phenylpropanoids*

Compounds like verbascoside (acteoside) have been extracted from *Lantana camara,* which possess antioxidant and anti-inflammatory action (Alipieva et al. 2014).

The potential synergistic activity of these diverse families of phytochemicals is primarily responsible for the holistic therapeutic effects seen in traditional medicine, but modern pharmacology is interested in identifying all contributors to guide their use more precisely.

8.4   The Double-Edged Sword: Toxicity and Safety Considerations

The significant therapeutic potential of *Lantana camara* comes with serious toxicity. This duality is the most significant attribute and warrants careful scientific investigation before any widespread use in medicine (Gamil et al. 2025).

8.4.1   *Lantadene Toxicity: The Primary Culprit*

The principal toxic compounds in *Lantana camara* are the group of pentacyclic triterpenoids, referred to as the lantadenes and especially lantadene A and B. The risk of toxicity is most significantly related to the level of concentration of the lantadenes in the leaves (young leaves) and unripe berries (Ramírez et al. 2025).

- *Hepatotoxicity:* The consumption of large amounts of lantadene (especially by grazing animals, for example, cattle, sheep, and goats) leads to severe liver damage or hepatotoxicity. The toxins cause cholestatic activity (bile duct obstruction) and hepatocellular necrosis (Pass et al. 1979).

- *Photosensitization:* The liver failure causes the excretion of phylloerythrin (chlorophyll metabolite) to be impaired, and the compound subsequently begins to accumulate in the body leading to photosensitivity which results in severe skin lesions, edema, and inflammation of unpigmented areas (sunburn) when exposed to sunlight (Sharma et al. 2012; Rasool et al. 2024). Organisms that cannot "tan" are very commonly affected. This is perhaps the most debilitating aspect of the poisonous effects in livestock.
- *Gastrointestinal and Renal Effects:* Lantadenes can cause gastrointestinal irritation, which manifests as anorexia, vomiting, and diarrhoea, and excessive quantities can impact kidney activity, leading to damage (Ramírez et al. 2025).

8.4.2 Human Toxicity

While severe cases of *Lantana camara* poisoning in humans are less common than in livestock, they do exist, particularly in children who may eat the attractive, bright-colored berries (Carstairs et al. 2010). Human cases tend to show a pattern of gastrointestinal upset (nausea, vomiting, diarrhoea), dilated pupils, and more extreme neurological variations or liver failure in the more extreme cases. Some have reported fatalities, but these are rare from severe ingestion (Sharma et al. 2007).

8.5 Future Directions

The transition from traditional knowledge to scientific validation has a long way to go but has provided some interesting insights regarding *Lantana camara*.

- *Standardization and Quality Control:* For *Lantana camara* to be considered a viable source of therapeutic agents, rigorous standardization of extracts, ensuring consistent phytochemical profiles and safety, is crucial. For *Lantana camara* to be considered a source of therapeutic agents, it will be imperative to standardize extracts in order to create future batches that promise similar phytochemical profiles and safety.
- *Isolation of Bioactive Compounds:* More research is needed to isolate and characterize individual components that are bioactive, especially those with some anti-inflammatory, antimicrobial, and antimalarial activities, and moreover to lessen or oxidize the toxic compounds.
- *Clinical Trials:* High-quality clinical trials are important for evidence of efficacy and safety of *Lantana camara* extracts in humans for various diseases. Within clinical trials, dose-response studies and long-term safety studies would be done.
- *Sustainable Sourcing and Bioprospecting:* Because *Lantana camara* is an invasive species, its use for medicinal purposes might be ecologically beneficial since it will be harvested. Bioprospecting studies should look for tolerable genotypes

characterized by high concentrations of beneficial compounds and low levels of toxins.

- *Mechanism of Action Studies:* The potential for designing new drugs will be determined by our understanding of the molecular mechanisms behind their pharmacological activities.
- *Targeted Drug Development:* There is potential for *Lantana camara* as a source for new drugs with uses as anti-inflammatory, antimicrobial, or antimalarial agents and can use either synthesis of the analogues or additional approaches for drugs using the active compounds.

8.6   Conclusion

Lantana camara is a true paradoxical plant in being ecologically infamous as an invasive plant but historically ethnomedically accepted and now being evaluated for pharmacological potential. The number of bioactive compounds, especially triterpenoids, flavonoids, oils, etc., that we are finding in many species of *Lantana camara* suggests it is possible that the genus may be heralded for drug discovery to manage a wide range of health concerns, potentially including pathogen management, inflammation, and cancers. However, the total toxicity owed to lantadenes can't be ignored. A transition from traditional empirical medicine to modern therapeutics must remain vigilant about scientific rigor. Future avenues of research should appraise: (i) advanced phytochemical profiling, (ii) mechanism-based drug discovery, (iii) targeted isolation and derivatization, robust toxicological and clinical trials, and (iv) sustainable bioprospecting. *Lantana camara* ultimately stands as a fascinating manifestation of nature's somewhat appalling ability to harm and heal. If science were to disentangle the layers of complexities, we may see *Lantana camara* change from an ecological liability to a "medicinal warrior" that could positively impact global health, while also recognizing the overwhelming power it possesses.

References

Alipieva K, Korkina L, Orhan IE, Georgiev MI. Verbascoside – a review of its occurrence, (bio) synthesis and pharmacological significance. Biotechnol Adv. 2014;32(6):1065–76. https://doi. org/10.1016/j.biotechadv.2014.07.001.

Al-Snafi A. Chemical constituents and pharmacological activities of *Lantana camara* – a review. Asian J Pharm Clin Res. 2019;12(12):10–20. https://doi.org/10.22159/ajpcr.2019. v12i12.35662.

Bishayee A, Ahmed S, Brankov N, Perloff M. Triterpenoids as potential agents for the chemoprevention and therapy of breast cancer. Front Biosci. 2011;16(3):980–96. https://doi. org/10.2741/3730.

Carstairs SD, Luk JY, Tomaszewski CA, Cantrell FL. Ingestion of *Lantana camara* is not associated with significant effects in children. Pediatrics. 2010;126(6):1585–8. https://doi.org/10.1542/peds.2010-1669.

El-Banna AA, Darwish RS, Ghareeb DA, Yassin AM, Abdulmalek SA, Dawood HM. Metabolic profiling of *Lantana camara* L. using UPLC-MS/MS and revealing its inflammation-related targets using network pharmacology-based and molecular docking analyses. Sci Rep. 2022;12:14828. https://doi.org/10.1038/s41598-022-19137-0.

Elekofehinti OO, Iwaloye O, Olawale F, Ariyo EO. Saponins in cancer treatment: current progress and future prospects. Pathophysiology. 2021;28(2):250–72. https://doi.org/10.3390/pathophysiology28020017.

Gamil NM, Elsayed HA, Hamed RM, Salah ET, Ahmed AM, Mostafa HA, El-Shiekh RA, Abou-Hussein D. Insights from herb interactions studies: a foundational report for integrative medicine. Future J Pharm Sci. 2025;11:46. https://doi.org/10.1186/s43094-025-00794-7.

Jabeen A, Khan MA, Ahmad M, Zafar M, Ahmad F. Indigenous uses of economically important flora of Margallah hills national park, Islamabad, Pakistan. Afr J Biotechnol. 2009;8(5):763–84.

Kakade RD, Lasure U, Kedare H, Bhingare P, Pokale S. Extraction, isolation, and purification of bioactive compounds from Lantana camara leaves: a comprehensive review. Res J Sci Technol. 2025;17(2):177–82. https://doi.org/10.52711/2349-2988.2025.00025.

Kato-Noguchi H, Kato M. Compounds involved in the invasive characteristics of *Lantana camara*. Molecules. 2025;30(2):411. https://doi.org/10.3390/molecules30020411.

Khan M, Mahmood A, Alkhathlan HZ. Characterization of leaves and flowers volatile constituents of Lantana camara growing in central region of Saudi Arabia. Arab J Chem. 2016;9(6):764–74. https://doi.org/10.1016/j.arabjc.2015.11.005.

Kumar R, Sharma R, Patil RD, Mal G, Kumar A, Patial V, Kumar P, Singh B. Sub-chronic toxicopathological study of lantadenes of *Lantana camara* weed in Guinea pigs. BMC Vet Res. 2018;14(1):129. https://doi.org/10.1186/s12917-018-1444-x.

Lin D, Xiao M, Zhao J, Li Z, Xing B, Li X, Kong M, Li L, Zhang Q, Liu Y, Chen H, Qin W, Wu H, Chen S. An overview of plant phenolic compounds and their importance in human nutrition and management of type 2 diabetes. Molecules. 2016;21(10):1374.

Nayak BS, Raju SS, Eversley M, Ramsubhag A. Evaluation of wound healing activity of *Lantana camara* L.- a preclinical study. Phytother Res. 2009;23(2):241–5. https://doi.org/10.1002/ptr.2599.

Pass MA, Seawright AA, Lamberton JA, Heath TJ. Lantadene a toxicity in sheep. A model for cholestasis. Pathology. 1979;11(1):89–94. https://doi.org/10.3109/00313027909063543.

Ramírez J, Armijos C, Espinosa-Ortega N, Castillo LN, Vidari G. Ethnobotany, phytochemistry, and biological activity of extracts and non-volatile compounds from Lantana camara L. and semisynthetic derivatives – an updated review. Molecules. 2025;30(4):851. https://doi.org/10.3390/molecules30040851.

Rasool F, Nizamani ZA, Ahmad KS, Parveen F, Khan SA, Sabir N. An appraisal of traditional knowledge of plant poisoning of livestock and its validation through acute toxicity assay in rats. Front Pharmacol. 2024;15:1328133. https://doi.org/10.3389/fphar.2024.1328133.

Sagar L, Sehgal R, Ojha S. Evaluation of antimotility effect of *Lantana camara* L. var. acuelata constituents on neostigmine induced gastrointestinal transit in mice. BMC Complement Altern Med. 2005;5:18. https://doi.org/10.1186/1472-6882-5-18.

Sharma OP, Sharma S, Pattabhi V, Mahato SB, Sharma PD. A review of the hepatotoxic plant *Lantana camara*. Crit Rev Toxicol. 2007;37(4):313–52. https://doi.org/10.1080/10408440601177863.

Sharma N, Lal K, et al. Secondary photosensitization due to *Lantana* poisoning in buffalo. J Anim Res. 2012;1:35.

Shen M, Yuan L, Zhang J, Wang X, Zhang M, Li H, Jing Y, Zeng F, Xie J. Phytosterols: physiological functions and potential application. Foods. 2024;13(11):1754. https://doi.org/10.3390/foods13111754.

Tadesse E, Engidawork E, Nedi T, Mengistu G. Evaluation of the anti-diarrheal activity of the aqueous stem extract of *Lantana camara* Linn (Verbenaceae) in mice. BMC Complement Altern Med. 2017;17(1):190. https://doi.org/10.1186/s12906-017-1696-1.

Wu P, Song Z, Wang X, Li Y, Li Y, Cui J, Tuerhong M, Jin D, Abudukeremu M, Lee D, Xu J, Guo Y. Bioactive triterpenoids from *Lantana camara* showing anti-inflammatory activities in vitro and in vivo. Bioorg Chem. 2020;101:104004. https://doi.org/10.1016/j.bioorg.2020.104004.

Zahra M, Abrahamse H, George BP. Flavonoids: antioxidant powerhouses and their role in nano-medicine. Antioxidants (Basel). 2024;13(8):922. https://doi.org/10.3390/antiox13080922.

Chapter 9
Turning the Tide: Leveraging *Lantana camara* for Sustainable Environmental Applications

Abstract *Lantana camara* is considered a noxious weed and is oftentimes stigmatized in literature and media. The real or perceived stigma surrounding *Lantana camara* could cloud the real opportunity for bioprospecting it provides. While much of the previous research about *Lantana camara* has focused on its negative impacts on the environment, this chapter identifies the major secondary metabolite diversity, terpenoids, flavonoids, and alkaloids of *Lantana camara* and showcases its previously documented biological activities that are potentially valuable to foster ecological solutions. The chapter further explores specific environmental applications, such as phytoremediation of heavy metals and organic pollutants, development of eco-friendly bioherbicides and biopesticides, and as an input in bioenergy and composting. Selected case studies from around the world are included, illustrating ongoing research, innovative applications, and pragmatic uses of *Lantana camara* to encourage sustainable environmental management. This chapter argues for a shift in how researchers and practitioners view *Lantana camara* and for a pivot away from it being a "problem species" to bioprospecting opportunity for novel and green technologies.

9.1 Introduction: The Invasive Paradox

Lantana camara is one of the 100 most problematic and invasive alien species in the world due to its high adaptability, fast rates of growth and seed production, and the ability to produce allelochemicals (IUCN 2009). The ecological implications of *Lantana camara* are extensive, resulting in loss of biodiversity, alteration of nutrient cycling, and inhibition of native plant restoration and further reduces agricultural and forestry economic productivity (Negi et al. 2019).

At the same time, *Lantana camara* is also justified for its bioprospecting potential. Invasive species become weedy in particular because of their complex chemical defences and proficient use of resources (Kato-Noguchi and Kurniadie 2021). Seeing that these traits exist, this means that they could also be exploitable as a source of unique bioactive compounds. In this chapter, *Lantana camara* is considered not as just a pest of an ecosystem but a bioproducts source. The phytochemistry of *Lantana camara* is explored with its possible uses in environmental applications in phytoremediation, biopesticides/herbicides, and bioenergy, providing global case studies to demonstrate the importance of looking at invasive species from this perspective (Raj 2017).

9.2 The Dual Nature of *Lantana camara*: Threat and Resource

9.2.1 *Ecological and Economic Impacts as an Invasive Species*

Lantana camara grows dense, impenetrable thickets that will outcompete the local vegetation, reducing light levels and shade. Changes in microclimates through shading can limit species diversity and drive down populations of indigenous flora and fauna (Wild 2019). The leaves and fruit of *Lantana camara* are toxic to livestock, so economic losses are seen by livestock farmers; it can also act as a reservoir of crop pests and diseases (Ntalo et al. 2022). Control measures involve culpable factors like mechanical removal, and many chemical herbicides, and also biological control agents. Control measures are expensive, require huge amounts of labour cost and energy inputs, have varied success, and often yield their own environmental impacts.

9.2.2 *Chemical Richness: A Foundation for Bioprospecting*

Despite its negative ecological footprint, *Lantana camara* is an established source of immense primary metabolites. The invasiveness of *Lantana camara* is partly because of its own chemical load, which informs the way it interacts with herbivores, pathogens, and competing plants (Kumar et al. 2024). There is a complex array of compounds contained in *Lantana camara*, in the leaves, stems, flowers, and fruits, that increase its value as a potential target for chemical analysis and subsequent resource exploitation. This chemical diversity can only form the springboard for bioprospecting of *Lantana camara*.

9.3 Environmental Applications: Bioprospecting Avenues

The high chemical diversity and growth potential of *Lantana camara* suggest that this plant can be valuable for many aspects of environmental remediation.

9.3.1 Phytoremediation

One area in which *Lantana camara* has considerable potential is phytoremediation, defined as the use of plants to remediate/remove, degrade, or contain environmental contaminants; *Lantana camara* has excellent potential in phytoremediation with its high biomass accumulation, extensive root system, and ability to grow well under very stressful environmental conditions Pandey et al. 2016).

- *Heavy Metal Uptake*: Experimental research has shown that *Lantana camara* has the ability to uptake heavy metals including lead (Pb), cadmium (Cd), chromium (Cr), nickel (Ni), and zinc (Zn) found in contaminated soils and waste water. The plant can accumulate these metals within the roots (or root) and shoots (or stem) and uses them as a method of phytoextraction.
- *Degradation of Organic Pollution: Lantana camara* may also be able to degrade some organic molecules, but at this time, this remains an area for further investigation.

9.3.2 Bioherbicides and Biopesticides

The allelopathic effects of *Lantana camara* have been mainly attributed to triterpenoids and phenolics, providing a source for natural herbicides. Extract from *Lantana camara* was reported to inhibit seed germination and plant growth of subsequent successor species (Mishra 2011; Gindri et al. 2020). Alkaloids, essential oils, and other secondary metabolites, gathered from different plant parts of *Lantana camara*, exhibited insecticidal, larvicidal, nematicidal, and repellent properties against agricultural insect pests and disease vectors. *Lantana camara* can provide a biopesticide.

9.3.3 Biofuel and Bioenergy Production

The rapid growth and high biomass yield of *Lantana camara* also warrant consideration as a biomass feedstock for bioenergy. The plant can be utilized in solid biofuels (briquettes), biogas production via anaerobic digestion, or as a liquid biofuel (bioethanol, biodiesel) through numerous thermochemical or biochemical conversion processes (Koricho et al. 2017; Sinha et al. 2021; Inayat et al. 2022). Utilization of *Lantana camara* for bioenergy will not only provide with an alternative renewable energy source, but it can also help with the environmental management of this invasive species.

9.3.4   Compost and Biofertilizer

The plentiful biomass of *Lantana camara* can be turned into organic compost or biofertilizer. Composting will help to neutralize the plant's toxic compounds through microbial degradation while providing a valuable resource for the growing of soil amendments rich in nutrients. This approach to degradation offers a solution for both the management of waste and improving the fertility of soil. This is especially true for soils that are degraded and where *Lantana camara* can be helped to degrade (Rawat and Suthar 2014).

9.3.5   Soil Rehabilitation and Erosion Control

Lantana camara, a well-known colonizer of degraded soils, may be used for soil rehabilitation purposes. It has valuable attributes to assist soil erosion and soil stabilization due to its deep-rooting habit, especially in locations of past deforestation, overgrazing, and mining activity. *Lantana camara*'s rapid growth and vigorous regeneration make it a suitable species and ground cover which will reduce the soil cover photo exposed to geologic erosion with wind and water.

Lantana camara also may be utilized to improve soil fertility. Research has shown that *Lantana camara* can act as a mulch to improve soil fertility through decomposition and organic matter addition in the Indian Himalayan mountain regions (Kumar et al. 2009). *Lantana camara* may be used to restore soil degradation in rural areas by reestablishing mulch layers and organic matter (especially in tropical housing areas where soil erosion could be an issue). Proper consideration of the management of lantana is necessary for restored areas, as it could successfully invade and outcompete established native species while destabilizing the ecological balance.

9.3.6   Agroforestry and Land Restoration

Lantana camara can be used in a multipurpose land restoration project within an agroforestry system. Since *Lantana camara* can tolerate and thrive in poor soil, it may make it suitable for re-vegetating degraded land, particularly land that has been cleared through deforestation. In these agroecosystems, *Lantana camara* can provide benefits including the following: controlling soil erosion, supplying organic matter for nutrient enrichment of soil, as well as produce fuel wood and timber. Furthermore, it is fast-growing species and can also provide economic benefits with harvested biomass production.

Although the invasive aspects of *Lantana camara* must be managed, the agroforestry system can provide manage land in a more sustainable way. By having *Lantana camara* in mixed-species, agroforestry systems, it provides the opportunity to diversify production and improve ecosystem services, such as carbon stocks, water infiltration and retention, and conservation of biodiversity.

9.3.7 Carbon Sequestration and Climate Change Mitigation

Lantana camara can aid the mitigation of climate change via carbon sequestration. *Lantana camara* is a fast-growing plant, and when it grows, it utilizes huge amounts of carbon dioxide from the atmosphere and helps reduce greenhouse gas emissions. Moreover, it can grow in marginal and degraded soils and therefore be used to rehabilitate land that wouldn't be utilized for other carbon sequestration systems (Zhang et al. 2014).

In the context of agroforestry, *Lantana camara* may also be used as part of a carbon offset scheme, as its biomass will contribute to soil organic carbon stocks (Walker et al. 2018). However, it is important to ensure that any use of lantana does not result in the displacement of native species or disturbance of its local ecosystem, as clearing to eliminate its establishment may further undermine the intent of the carbon offset strategy overall (Lone et al. 2025).

9.4 Case Studies

Research and practical applications worldwide reveal expanding bioprospecting of *Lantana camara*:

9.4.1 Phytoremediation of Heavy Metals

In India, *Lantana camara* is a common invasive species, and it has been studied for its phytoremediation ability. It has been shown through experimentation to be capable of accumulating various heavy metals (Cd, Pb, Cu, Cr, Ni) in its roots and shoots, suggesting it has the potential to be utilized in areas of industrial pollution for phytoextraction, with the exception of Mn and Pb (Pandey et al. 2016). This is not only a way to clean soil but also a way to manage *Lantana camara*'s large biomass.

9.4.2 Bioherbicidal Potential

Recent years have witnessed agriculture has globally suffered from weed invasion; *Lantana camara* is, itself, a highly significant environmental weed. Scientists have investigated the allelopathic potential of *Lantana camara* extracts seen in this area as biopesticides or natural herbicides. Work has shown that aqueous extracts of *Lantana camara* leaves drastically reduced common agricultural weeds such as *Parthenium hysterophorus* (Mishra 2011) and *Bidens pilosa* (Gindri et al. 2020). These two research studies could be a launching point to develop green bioherbicides that could reduce our reliance on synthetic chemical herbicides while also providing management options for *Lantana camara*.

9.4.3 Bioenergy and Compost Production

Several countries dealing with *Lantana camara* infestation have considered it as a biomass feedstock. Some projects have studied its conversion from biomass into solid fuel briquettes for industrial and domestic heating, with good calorific value (Chongloi et al. 2024). Several studies have also looked at the co-composting of *Lantana camara* with other organic wastes, which successfully degraded its toxic compounds and produced nutrient-rich compost that is agriculturally acceptable compost (Rawat and Suthar 2014). These projects offer integrated solutions for invasive species management and sustainable practices.

9.5 Challenges and Future Perspectives

Despite some attractive opportunities, there are a number of barriers to overcome in order to achieve sustainable bioprospecting of *Lantana camara*. The occasional toxicities of certain compounds (e.g., lantadenes) must not be overlooked, particularly if biomass is being repurposed for agricultural use or as animal fodder. Conversely, the extraction and purification of specific bioactive compounds should be done in a sustainable and cost-effective manner; protocols successfully used in the lab must be developed, which can also be scaled up for a sustainable and cost-effective extraction and purification on the industrial scale.

Future research should consider the following areas of interest:

- *Targeted Bioprospecting:* This would involve identifying and extracting specific high-value compounds for pharmaceutical, agricultural or industrial uses.
- *Integrated Management Strategies:* This would explore a "whole-ecosystem approach," where removal of *Lantana camara* from the landscape and valorizing the species (through a variety of environmental benefits) could happen simultaneously.

- *Genetic and Omics Studies:* This would evaluate the genetic basis of invasiveness and secondary metabolites, in order to move from sedentary bioprospecting.
- *Life Cycle Assessment:* This would evaluate the carbon footprint associated with *L. camara*-based product, to ensure that the products are in fact sustainable.

9.6 Conclusion

A classic example of an alien species in an interstate dilemma: a dreaded ecological threat versus a hefty reservoir of bioactive compounds. The change in perception from the eradicative mindset toward solid and strategic bioprospecting embraces a sustainable approach for managing this ubiquitous weed. *Lantana camara* stands useful as a renewable bioresource for phytoremediation needs, bioherbicides, and biopesticides, among many other things, including bioenergy and compost production. Global case studies testify to the possibilities and implementations ongoing for the utilization of this plant. On the grounds of modern approaches in R&D, *Lantana camara* can be turned from a problem into the solution for sustainable environment and a new set of green technologies.

References

Chongloi V, Phukan M, Bora P. Miscellaneous prospects of invasive *Lantana camara* biomass – a standpoint on bioenergy generation and value addition. Environ Sci Pollut Res. 2024;31:59041–57. https://doi.org/10.1007/s11356-024-35042-7.

Gindri DM, Coelho CMM, Uarrota VG, Rebelo AM. Herbicidal bioactivity of natural compounds from *Lantana camara* on the germination and seedling growth of Bidens pilosa. Pesquisa Agropecuária Tropical. 2020;50:57746.

Inayat A, Ahmed A, Tariq R, Waris A, Jamil F, Ahmed SF, Ghenai C, Park YK. Techno-economical evaluation of bio-oil production via biomass fast pyrolysis process: a review. Front Energy Res. 2022;9:770355. https://doi.org/10.3389/fenrg.2021.770355.

IUCN. Invasive Species Specialist Group, 100 World's Worst Invasive Alien Species. http://www.issg.org/booklet (2009)

Kato-Noguchi H, Kurniadie D. Allelopathy of *Lantana camara* as an invasive plant. Plants. 2021;10(5):1028. https://doi.org/10.3390/plants10051028.

Koricho SA, Leta S, Soromessa T, Khan MM. Fuel briquette potential of *Lantana camara* L. weed species and its implications for Weed Management and recovery of renewable energy sources in Ethiopia. J Environ Sci Toxicol Food Technol. 2017;11:41–51.

Kumar P, Pant M, Negi GCS. Soil physico-chemical properties and crop yield improvement following *Lantana* mulching and reduced tillage in rainfed croplands in the Indian Himalayan mountains. J Sustain Agric. 2009;33:636–57.

Kumar R, Guleria N, Deeksha MG, Kumari N, Kumar R, Jha AK, Parmar N, Ganguly P, de Aguiar Andrade EH, Ferreira OO, de Oliveira MS. From an invasive weed to an insecticidal agent: exploring the potential of *Lantana camara* in insect management strategies – a review. Int J Mol Sci. 2024;25(23):12788. https://doi.org/10.3390/ijms252312788.

Lone PA, Kothandaraman S, Dar JA, Hakeem KR, Khan ML. Invasive shrub (*Lantana camara* L.) alters the tree diversity and ecosystem-level carbon pools in tropical forests of Central India. Front Glob Change. 2025;8:1412130. https://doi.org/10.3389/ffgc.2025.1412130.

Mishra A. Studies of allelopathic effect of *Lantana camara* aqueous leaf extract on growth of Parthenium hysterophorus in flowering stage. Indian Jo Appl Res. 2011;4:33–5. https://doi.org/10.15373/2249555X/June2014/9.

Negi GCS, Sharma S, Vishvakarma SCR, Samant SS, Maikhuri RK, Prasad RC, Palni LMS. Ecology and use of *Lantana camara* in India. Bot Rev. 2019;85:109–30. https://doi.org/10.1007/s12229-019-09209-8.

Ntalo M, Ravhuhali KE, Moyo B, Hawu O, Msiza NH. *Lantana camara*: poisonous species and a potential browse species for goats in Southern Africa- a review. Sustainability. 2022;14(2):751. https://doi.org/10.3390/su14020751.

Pandey SK, Bhattacharya T, Chakraborty S. Metal phytoremediation potential of naturally growing plants on fly ash dumpsite of Patratu thermal power station, Jharkhand, India. Int J Phytoremediation. 2016;18(1):87–93.

Raj S. Preliminary phytochemical screening of *Lantana camara*, L., a major invasive species of Kerala, using different solvents. Ann Plant Sci. 2017;6:1794. https://doi.org/10.21746/aps.2017.6.11.13.

Rawat I, Suthar S. Composting of tropical toxic weed *Lantana camera* L. biomass and its suitability for agronomic applications. Compost Sci Util. 2014;22:105–15. https://doi.org/10.1080/1065657X.2014.895455.

Sinha D, Banerjee S, Mandal S, Basu A, Banerjee A, Balachandran S, Mandal NC, Chaudhury S. Enhanced biogas production from *Lantana camara* via bioaugmentation of cellulolytic bacteria. Bioresour Technol. 2021;340:125652. https://doi.org/10.1016/j.biortech.2021.125652.

Walker SM, Pearson TRH, Casarim FM, Harris N, Petrova S, Grais A, Swails E, Netzer M, Goslee KM, Brown S, Sidman G. Standard operating procedures for terrestrial carbon measurement: version 2018. Winrock International; 2018.

Wild C. The dynamics of *Lantana Camara* (L.) invasion of subtropical rainforest in southeast Queensland; 2019. https://doi.org/10.25904/1912/293.

Zhang Q, Zhang Y, Peng S, Zobel K. Climate warming may facilitate invasion of the exotic shrub *Lantana camara*. PLoS One. 2014;9(9):e105500. https://doi.org/10.1371/journal.pone.0105500.

Chapter 10
The Hidden Potential: *Lantana camara* for Value-Added Products

Abstract This chapter explores the dual nature of *Lantana camara*, a problematic invasive weed, as an opportunity to utilize value-added products. As we highlighted, *Lantana camara* poses a significant ecological threat; however, we documented its application in the natural dye and pulp and paper industries. Phytochemical analysis supported that *Lantana camara* is rich in flavonoids, which results in many types of eco-friendly and sustainable textile dyes. Its fibrous stems represent a potential non-wood-based raw material that would be suitable for pulp in many different grades of paper. Overall, these various applications for *Lantana camara* provide many benefits, including a sustainable alternative for raw materials, infrastructure for local rural economies, etc. Further research is needed to address the inconsistencies in color or dyeing as well as efficiency in pulping. Overall, the use of *Lantana camara* in either of these industries (dyes or pulp) allows land managers the opportunity to sustainably manage an invasive species while finding some potential economic use.

10.1 Introduction

Lantana camara, a highly invasive species, poses severe environmental and economic problems throughout many tropical and subtropical areas globally. This species is proliferating rapidly and is highly allelopathic and toxic to livestock, costing significant amounts of money for management (Negi et al. 2019). Nonetheless, the same reasons why *Lantana camara* is problematic as a weed (vigorous growth, lot of biomass, and a unique chemical pathway) also show promise for significant value addition. A shift away from eradication, focusing instead on economically possible uses of *Lantana camara* biomass, may provide a longer-term feasible resolution of an invasiveness issue while also providing income and local industry (Kumar and Singh 2020).

This chapter will discuss several ways of changing *Lantana camara* from a bad weed to a potential source of value-added products. The possibilities to be reviewed for their potential in natural dyes, pulp and paper, bioenergy, and other novel uses,

M. A. Dervash et al., *Lantana Camara*, SpringerBriefs in Plant Science,
https://doi.org/10.1007/978-3-032-03837-1_10

scientific rationale, and reflections will also be highlighted for each. The goal is to demonstrate how redirecting this pervasive plant can meaningfully aid environmental management, resource utilization, and economic development.

10.2   *Lantana camara* as a Source of Natural Dyes

The bright colors of the *Lantana camara* flowers and the many phenolic compounds found in the foliage and stems suggest the potential of this plant as a source of natural dye. The emerging interest in more sustainable products—coupled with a general lack of interest in synthetic dyes by industry as a result of the associated risk to health and the environment—illustrates the potential for the use of natural dyes from such invasive weeds (Datta et al. 2022).

10.2.1   *Phytochemical Basis for Dye Potential*

Natural color compounds in vegetation are primarily pigments. Among them are flavonoids, carotenoids, and tannins. *Lantana camara* contains the following pigments (Datta et al. 2022):

- *Flavonoids:* This is a large group of polyphenolic compounds, typically associated with yellow, orange, and red pigments in a plant. *Lantana camara* leaves and flowers contain several flavonoids (Andersen and Markham 2005).
- *Tannins:* Tannins, which occur in bark and foliage, are known to have mordanting properties, which will yield brown, black, or gray shades (Datta et al. 2022).
- *Carotenoids:* Primarily responsible for yellow and orange colors, presumably in the flower (Datta et al. 2022).
- *Chlorophyll:* In fresh green leaves, chlorophyll is quite susceptible to degradation during processing, which may permit pigments beneath it to show or yield a green-brown color (Datta et al. 2022).

The particular hue and color intensity of dye extracted from *Lantana camara* depend on the following terms:

- *Plant part used:* Flowers, leaves, and stems may produce different colors. The multicolored flowers alone show that there are multiple pigments that can be extracted (Rawat et al. 2004).
- *Extraction method:* Water extraction (decoction), alkali extraction, or solvent extraction will each produce different types and concentrations of pigments (Rawat et al. 2004).
- *Mordants:* Natural dyes typically involve the use of a mordant (e.g., alum, iron salts, copper salts, etc.), which can fix the dye on the fabric, enhance the intensity of the color, and enhance wash light fastness. Mordants can also shift the resulting color (Rawat et al. 2004).

- *pH of the dye bath:* The pH of the dye bath can affect/change the color of the extract prior to dyeing (Rawat et al. 2004).

10.2.2 Traditional and Experimental Dyeing Applications

Not having been a dye plant of note in many regions historically, however, modern research has been investigating its potential.

- *Flower Extracts:* The bright orange and yellow flowers may provide yellow to orange dyes when applied to natural fibers such as cotton, wool, and silk. In addition to yellow or orange, using different mordants can produce variations of brown, green, or even reddish brown (Datta et al. 2022; Rawat et al. 2004).
- *Leaf Extracts:* Leaves, rich in flavonoids and chlorophyll, can produce yellow-green to brownish-yellow dyes (Datta et al. 2022; Rawat et al. 2004).
- *Whole Plant Extracts:* The leaf extracts produce a yellow-green to brownish-yellow color with a significant number of flavonoids and high chlorophyll content (Datta et al. 2022; Rawat et al. 2004).

10.2.3 Advantages and Challenges

Advantages
- *Sustainable Sourcing:* An invasive species is managed through the beneficial use of the dye flooding the marketplace.
- *Eco-Friendly Alternative:* Using a natural dye helps to lower the reliance on synthetic dyes, which are derived from petrochemicals and thus positively affect the environment and our health (Dhanush et al. 2020).
- *Renewable Resource:* This species grows rapidly and can always produce biomass if planted, theoretically producing dye indefinitely.
- *Local Employment:* The harvesters and processers are potential employment in rural communities.

Challenges
- *Color Fastness:* Natural dye color fastness is often lower than synthetic counterparts for wash and light fastness, with natural dyes requiring specific techniques of mordanting in order to be semi-permanent.
- *Consistency:* Color yield and shade can vary significantly depending on plant maturity, growing conditions, and extraction protocols.
- *Scaling Up:* Scaling from small lab-scale extraction to industrial production takes optimized cost and time-efficient protocols.
- *Toxicity Concerns:* While exposure to dyes applied externally does not raise many flags, once larger quantities of biomass are included, other concerns, such as residual toxins present in the leftover biomass or dealing with the large quantities of materials, may arise as problems.

10.3   *Lantana camara* for Pulp and Paper Production

The fibrous quality of *Lantana camara* stems and its high biomass yield suggest it is a high-potential candidate for pulp and paper manufacturing, especially where wood fiber is scarce or where alternative fiber sources are needed (Bhodiwal et al. 2024).

10.3.1   Fiber Characteristics and Chemical Composition

The key characteristics for paper production include the fiber length, cellulose, lignin, and hemicellulose content.

- *Cellulose:* The main part of the wall that gives strength and adds to the structure of the paper. *Lantana camara* has a respectable cellulose content, comparable to some non-wood sources (Bhodiwal et al. 2022).
- *Lignin:* A binding agent, but must be removed during the pulping process. The amount of lignin will determine how difficult, expensive, or energy intensive is pulping process (Bhodiwal et al. 2022).
- *Hemicellulose:* Imparts the strength and formation of the paper (Bhodiwal et al. 2022).
- *Fiber Length:* Longer fibers are generally better, as they produce stronger paper. *Lantana camara* fiber is generally shorter than wood fibers, but comparable with other non-wood fibers such as bagasse or rice straw, and thus can be used for some grades of paper (Bhodiwal et al. 2022).

Numerous studies have investigated the pulping of *Lantana camara*:

- *Sulphate (Kraft) Pulping:* Research indicated that *Lantana camara* could be pulped with the Kraft method, yielding pulp that was acceptable for writing and printing paper; the only concern may be bleachability due to lignin (Iglesias et al. 2019).
- *Soda Pulping:* A method that is alkaline and more environmentally safe than Kraft pulping because there are no sulphur compounds (Bhodiwal et al. 2024).
- *Mechanical Pulping:* This is not very common because the lignin content is very high, and therefore, this would be of lower strength (Bhodiwal et al. 2024).

10.3.2   Potential Paper Products

- *Writing and Printing Paper: Lantana camara*, with pulping and bleaching, could provide pulp for potential use for writing and printing papers, and in cases, it may be blended with wood pulp for improved pulp properties (Pydimalla and Reddy 2020).

- *Packaging Boards:* Its fiber may contribute to coarser paper products, as in corrugated medium, linerboard, and other packaging materials (Pydimalla and Reddy 2020).
- *Specialty Papers:* Its unique fiber qualities may be suitable for some specialty papers (Pydimalla and Reddy 2020).

10.3.3 Advantages and Challenges

Advantages
- *Abundant Biomass:* The sheer volume of *Lantana camara* provides an abundant and renewable fiber source.
- *Fast Growth Rate:* Fast regeneration rates make it one of the most sustainable fiber sources without deforestation.
- *Alternative Fiber Source:* Using *Lantana camara* as an alternative source of fiber reduces the pressure on forest resources or provides a fiber solution in areas with limited timber resources.
- *Waste to Wealth:* It converts an environmental problem into a valuable resource.

Challenges
- *Pulping Efficiency:* The lignin content and unique fiber morphology may require optimized pulping conditions, with increased chemical consumption or energy input compared to traditional wood when pulping *Lantana camara*.
- *Paper Quality:* Due to the fiber length of intrinsic strength, it may not be suitable for full use properly nor for high quality without blending.
- *Collection and Transportation:* Collecting and transporting large volumes of *Lantana camara* biomass to a pulping facility in an efficient and economical manner may be difficult.
- *Toxic Residues:* Where residual lantadenes are remaining either in the pulp or wastewater, they have to be carefully managed or processed to avoid introducing contaminants to the environment.
- *Economic Viability:* It is important to determine if the cost of *Lantana camara* would be competitive with existing fiber sources.

10.4 Other Value-Added Products from *Lantana camara*

In addition to dyes and paper, *Lantana camara* biomass may be considered for many other value-added uses that are consistent with the principles of a circular economy.

10.4.1 Biofuel and Bioenergy

- *Briquetting/Pelleting:* The dried biomass can be compacted into briquettes or pellets, which provide a renewable solid fuel and key alternative for industrial and domestic heating applications so as to complement less sustainable fossil fuel usage (Koricho et al. 2017).
- *Biogas Production:* Anaerobic digestion of *Lantana camara* biomass can yield biogas (primarily methane), which can be used for the purpose of generating electricity or for direct heating (Sinha et al. 2021).
- *Bioethanol:* Properly pretreated, cellulose can be fermented to form bioethanol.
- *Bio-oil/Syngas:* The technique of pyrolysis or gasification of *Lantana camara* biomass can be used to make bio-oil, a liquid fuel, or syngas, a mixture of CO, H_2, and CO_2, which can be utilized as energy generation or different chemical feedstocks (Inayat et al. 2022).

10.4.2 Biopesticides and Insecticides

The essential oil and some extracts of *Lantana camara* have insecticidal properties and larvicidal capabilities, making it a viable source for:

- *Botanical Pesticides:* Reasonable natural pesticide made from *Lantana camara* will provide eco-friendly opportunities for agriculture and a means of using an alternative option compared to synthetic chemicals (Kumar et al. 2024).
- *Mosquito Repellents/Larvicides:* There is the potential for some extracts to be inserted in formulations to manage mosquito populations and facilitate in managing disease vectors (Sonter et al. 2024).

10.4.3 Compost and Organic Fertilizers

Although *Lantana camara* has allelopathic potential in its fresh state, *Lantana camara* biomass that has been processed (e.g., fully composted) can still be safely used as organic fertilizers or soil amendments. Composting should break down the toxicants and release nutrients back in the soil so that we can obtain good soil health and fertility (Yadav and Argaw 2016).

10.4.4 Pharmaceutical and Nutraceutical Applications

It is worthwhile to mention again that while the focus of the last chapter was isolation of particular bioactive compounds from *Lantana camara* for the development of anti-inflammatory, antimicrobial, or antimalarial drugs. The potential high-value target must be taken care that the toxicity risk is mitigated properly (Kakade et al. 2025).

10.5 Implementation Challenges

To effectively roll out value-added activities for *Lantana camara*, it is essential to take a multidisciplinary approach by bringing together evidence based scientific work with economic viability and community engagement (Priyanka et al. 2013).

- *Current Initiatives:* Pilot activities have commenced in a number of areas (including parts of India), which utilize *Lantana camara* to produce furniture, handicrafts, and briquettes. These provided insights into the potential of local active ingredients.
- *Technological Gaps:* Attempts to establish processes from the laboratory scale to industrial scale will encounter limitations around process optimization, economic viability, and waste management at scale.
- *Logistics and Supply Chain:* Ways for obtaining, processing, and shipping bulky *Lantana camara* biomass (as well as other invasive species) will be essential in order to make it economically viable.
- *Policy Support:* Government policy and guidelines that encourage the use of invasive species, provide subsidies for research into value addition, and support communities involved in local livelihoods will be important.
- *Community Participation:* Community involvement in the collection and early stages of processing will provide community livelihoods and ensure a sustainable supply via the harvest of raw material.

10.6 Conclusion

The transformation of rendering *Lantana camara* from being just an environmentally harmful invader to exploring its potential and discussing it as a global resource is a true example of waste valorization and sustainable development. There is much more research and development to be conducted, but the prospects of turning this invasive plant into natural dyes, pulp and paper, bioenergy, and other applications are hopeful.

Making value-added products from *Lantana camara* and using it as a resource have the potential to also be an environmentally sustainable way to manage an invasive weed, build economic development opportunities, and create new resources. By investing in green technologies, providing support for collaborative research, and developing supportive policies, we can tell the story of *Lantana camara* as a solution rather than a problem and help build a sustainable resource economy. This invasive plant, which is often viewed as the picture of ecological degradation, may soon be seen as a new picture of human ingenuity and developing a solution from a problem.

References

Andersen OM, Markham KR. Flavonoids: chemistry, biochemistry and applications. CRC Press; 2005.

Bhodiwal S, Agarwal R, Chauhan S. Evaluation for characterization of pulp and paper making from weed species *Lantana camara* L. Biospectra. 2022;17(2):187–92.

Bhodiwal S, Agarwal R, Chauhan S. A weed species Lantana camara L.: an alternative raw material in handmade paper making. TWIST. 2024;19(1):268–74.

Datta DB, Das D, Sarkar B, Majumdar A. *Lantana camara* flowers as a natural dye source for cotton fabrics. J Natu Fibers. 2022;20(1) https://doi.org/10.1080/15440478.2022.2159604.

Dhanush C, Naveen Kumar TD, Kiran Kumar BT, Sathish BN, Hareesh TS, Gajendra CV, Ashwath MN. Extraction and chemical characterization of leaf and flower dye from *Lantana camara* Linn. J Pharm Phytochem. 2020;9(3):842–51. https://doi.org/10.22271/phyto.2020.v9.i3n.11385.

Iglesias M, Gomez-Maldonado D, Peresin M, Via B. Comparison of Kraft and sulfite pulping processes and their effects on cellulose fibers and nanofibrillated cellulose properties: a review. For Prod J. 2019;70:10–21. https://doi.org/10.13073/FPJ-D19-00038.

Inayat A, Ahmed A, Tariq R, Waris A, Jamil F, Ahmed SF, Ghenai C, Park Y-K. Techno-economical evaluation of bio-oil production via biomass fast pyrolysis process: a review. Front Energy Res. 2022;9:770355. https://doi.org/10.3389/fenrg.2021.770355.

Kakade RD, Lasure U, Kedare H, Bhingare P, Pokale S. Extraction, isolation, and purification of bioactive compounds from *Lantana camara* leaves: a comprehensive review. Res J Sci Technol. 2025;17(2):177–82. https://doi.org/10.52711/2349-2988.2025.00025.

Koricho SA, Leta S, Soromessa T, Khan MM. Fuel briquette potential of *Lantana camara* L. weed species and its implications for weed management and recovery of renewable energy sources in Ethiopia. J Environ Sci Toxicol Food Technol. 2017;11:41–51.

Kumar RP, Singh JS. Invasive alien plant species: their impact on environment, ecosystem services and human health. Ecol Indic. 2020;111:106020. https://doi.org/10.1016/j.ecolind.2019.106020.

Kumar R, Guleria N, Deeksha MG, Kumari N, Kumar R, Jha AK, Parmar N, Ganguly P, de Aguiar Andrade EH, Ferreira OO, de Oliveira MS. From an invasive weed to an insecticidal agent: exploring the potential of Lantana camara in insect management strategies – a review. Int J Mol Sci. 2024;25(23):12788. https://doi.org/10.3390/ijms252312788.

Negi GCS, Sharma S, Vishvakarma SCR, Samant SS, Maikhuri RK, Prasad RC, Palni LMS. Ecology and use of *Lantana camara* in India. Bot Rev. 2019;85:109–30. https://doi.org/10.1007/s12229-019-09209-8.

Priyanka N, Shiju MV, Joshi PK. A framework for management of *Lantana camara* in India. Proc Int Acad Ecol Environ Sci. 2013;3:306–23.

Pydimalla M, Reddy K. Effect of pulping, bleaching and refining process on fibers for paper making – a review. Int J Eng Res Technol. 2020;9(12):310–6.

Rawat B, Jahan S, Sharma E, Yadav S. *Lantana* flowers – an ecofriendly natural dye for silk. Asian Textile J. 2004;13:89–91.

Sinha D, Banerjee S, Mandal S, Basu A, Banerjee A, Balachandran S, Mandal NC, Chaudhury S. Enhanced biogas production from *Lantana camara* via bioaugmentation of cellulolytic bacteria. Bioresour Technol. 2021;340:125652. https://doi.org/10.1016/j.biortech.2021.125652.

Sonter S, Dwivedi MK, Mishra S, Singh P, Kumar R, Park S, Jeon BH, Singh PK. In vitro larvicidal efficacy of *Lantana camara* essential oil and its nanoemulsion and enzyme inhibition kinetics against *anopheles culicifacies*. Scient Rep. 2024;14(1):16325. https://doi.org/10.1038/s41598-024-67148-w.

Yadav H, Argaw A. Biodegradation of *Lantana camara* using different animal manures and assessing its manurial value for organic farming. South Indian J Biol Sci. 2016;2:2454–4787. https://doi.org/10.22205/sijbs/2016/v2/i1/100344.

Chapter 11
Toxicology and Safety: Understanding and Mitigating the Risks of *Lantana camara*

Abstract This chapter thoroughly discusses the important toxicological properties of *Lantana camara*, a plant that is both notably invasive and has had traditional benefits. This chapter also details the negative effects that result from lantadene triterpenoids on livestock, specifically to the degree observed in regions, where serious liver damage and photosensitization cause economic impacts. Rarely will these symptoms occur in humans, but there are cases of gastrointestinal upset, and even rarer are systemic effects, most commonly related to accidental ingestion of unripe berries. The chapter emphasizes the value of detoxification strategies, including supportive care in poisoned animals and humans, as well as advanced methods including chemical methods and selective extraction methods to make Lantana biomass safe for valorization, and also realizing the value of these plants responsibly.

11.1 Introduction

Lantana camara is a very widespread plant throughout the tropical and sub-tropical regions and it is an enigma. Throughout the previous chapters while its sociocultural uses and the exploration of its use for the development of various value-added products have been explored, it is just as, if not more well regarded for its notorious toxicity (Sharma et al. 2007). The tendency for people to overlook *Lantana camara*'s toxic nature leads to the need to better understand the negative health impacts, especially with the livestock that regularly graze on unsustainable amounts of leaves or fledgling plants, and on humans through incidental ingestion (Ntalo et al. 2022). As farmers frequently suffer substantial economic loss as a result of *Lantana* poisoning, the need for mitigation and management is critical (Sharma et al. 1981). This chapter will address the toxicology and safety profile of *Lantana camara* in detail, including the specific toxic compounds, clinical signs of poisoning in livestock and man, and solid approaches to risk assessment of *Lantana camara* products and extracts. The viable detoxification approaches for responding to incidents of poisoning and rendering *Lantana camara* biomass safe for potential valorization are discussed. Together all these contents will give a better understanding of the

M. A. Dervash et al., *Lantana Camara*, SpringerBriefs in Plant Science,
https://doi.org/10.1007/978-3-032-03837-1_11

risks associated with and to effectively manage these risks for the sake of animal health and human safety, required for any utilization of this plant that is undoubtedly very prevalent.

11.2   Toxicity and Adverse Effects of *Lantana camara*

Lantadenes are a specific group of chemicals responsible for the toxicity of *Lantana camara* (Kato-Noguchi and Kato 2025).

11.2.1   Key Toxic Constituents: Lantadenes

The toxic agents in *Lantana camara* are principally pentacyclic triterpenoids, namely, lantadene A and lantadene B. Lantadenes are principally distributed in the leaves of the plant and focus on the most tender and fresh, young leaves (Sharma and Sharma 1989). Lantadenes are distributed in the stems of the plant and are also found in the undeveloped berry (unripe berries). The distribution and actual types of lantadene can vary a lot based on the plant genotype (this includes different varieties and chemotypes), environmental conditions (type of soil and weather), and the plant stage of maturity (Negi et al. 2019). These geographical, seasonal, environmental, and physiological factors contribute to the variability in tempered toxicity in infestations in certain areas.

11.2.2   Toxicity in Livestock

Livestock poisoning with *Lantana camara* is a global issue that represents serious economic losses at an individual farm level. Ruminants, specifically cattle, sheep, and goats, are the most sensitive to the effects of lantadenes (Sharma and Makkar 1981). When animals graze on *Lantana camara* foliage, especially during times of scarce forage, the lantadenes are absorbed from the gastrointestinal tract, and the primary toxic effect is experienced in the liver, where lantadenes cause hepatocellular damage, or damage to the liver cells. A major consequence of the hepatocellular damage is the direct impairment of bile flow, called *cholestasis*, which prevents the normal excretion of bile pigments, leading to excess leakage and resulting in bile pigments accumulating in the liver and blood (Kumar et al. 2018). Importantly, the cholestasis prevents the normal excretion of phylloerythrin, a breakdown product of chlorophyll that rumen microbes produce for all ruminants. Phylloerythrin will build up in the blood, allowing for skin absorption, and when animals have unpigmented or poorly pigmented areas of skin, the phylloerythrin reacts quickly under ultraviolet (UV) light to form reactive oxygen species that cause severe photosensitivity and severe damage to the skin (Rowe 1989; Pass 1986).

The clinical signs of lantadene toxicity, better known as *Lantana camara* poisoning, initially present as gastrointestinal and systemic signs (Kumar et al. 2018). Such signs include anorexia, lethargy, depression, constipation, decreased rumination, and, in lactating mammals, a significant decline in milk production. In very severe cases, rumen or omasum impaction can also occur. Liver signs are a primary feature of the poisoning, and jaundice (icterus) is common as bile pigments accumulate, leading to yellowing of the mucous membranes (eyes and gums). Evidence of liver damage would come from elevated liver enzymes (AST, GGT, ALP) (Day et al. 2003). The most definitive signs are those related to photosensitization of nonpigmented or sparsely haired skin exposed to sunlight. Signs of photosensitization include swelling (edema) of the muzzle, eyelids, ears, and sometimes the entire face, reddening of affected skin like a sunburn, thick, hardened skin like a sunburn (Kumar et al. 2018). The hair can fall out, and in many cases, exudation with crusting on the affected areas may appear. In severe cases, skin areas may become necrotic (destruction and death of tissue) and slough off, leading to open wounds that are highly susceptible to secondary bacterial infection. These skin lesions are generally associated with very severe pruritus (itching) that prompts animals to rub or scratch their skin and only damages it further (Shyamkumar et al. 2021). Animals that survive acute poisoning may experience chronic liver damage and long-term weight loss and poor thrift. The outcome for affected livestock varies with the amount of plant ingested, exposure duration, and veterinary treatment time; in the most severe cases, some animals may succumb, either from liver failure or due to complications associated with secondary infections (Shyamkumar et al. 2021). All postmortem examinations reveal that the liver will often be large, pale, or yellow, sometimes with fibrous changes in the chronic cases and in all cases with identified gallbladder distension (Shyamkumar et al. 2021). Additionally, in some cases, there could be edema and inflammation in the subcutaneous tissues beneath the areas of skin affected. Overall, these effects can lead to considerable negative economic effects for farmers within infested areas, not only from animal deaths but loss of production and veterinary costs (Shyamkumar et al. 2021).

11.2.3   Toxicity in Humans

While human poisoning cases involving *Lantana camara* are far less frequent than in livestock, they do happen, typically as a result of accidental consumption as opposed to intentional (Carstairs et al. 2010). Most frequently, the risk comes from the consumption of unripe berries, which are toxic, especially to young children who are attracted to the appearance of the berries. Contact with plant tissue will only cause a localized skin reaction that doesn't produce systemic effects in humans. The main concern with accidental ingestion is that the lantadenes cause chemical injury. Clinical signs associated with accidental ingestion of Lantana can vary in severity from mild to moderate, and while nausea, vomiting, abdominal pain, and diarrhea are commonly seen clinical signs and are usually self-limiting, there may be other mild signs such as dilated pupils and muscle weakness. Human poisoning can vary in severity and is dependent on the amount of plant material ingested, the state of ripeness of the berries (unripe berries are more toxic), and the person's age and health (Stegelmeier et al. 2013).

11.3 Risk Assessment of *Lantana camara* Products and Extracts

Given the toxicity of *Lantana camara*, any market development for value-added products (be they medicinal extracts, dyes, paper or other value adding processes) requires a very detailed and serious risk management process. The objective, of course, is to protect consumers as well as the environment throughout the process and the final product.

A key element of risk assessment is the chemical profiling and quantification of toxicity (Pour et al. 2011). In this case, it is critical to identify and quantify containment of lantadene(s) and other harmful chemicals (if any) not only within all parts of the *Lantana camara* plant being used but also in any product and extract being made from any and all parts of that plant. This means that it is essential to use advanced analytical technology to identify and quantify chemical, complex molecules (masses and structures). Technologies such as High Performance Liquid Chromatography (HPLC) and Liquid Chromatography-Mass Spectrometry (LC-MS/MS) have the ability to quantify compounds that may otherwise either not be detected or difficult to have accurate concentrations (quantity). Given the variability in plant compositions (even for the same plant), monitoring for levels of lantadene in raw material and finished products would need to be continuous. Lastly, it would be important to establish an acceptable daily intake (ADI) or safe levels of exposure for specific Lantana-derived products based on careful toxicological data.

After chemical characterization has been conducted, toxicity testing of any extracts and products is necessary before any *Lantana* product can be deemed safe for human or animal exposure (Pour et al. 2011). This may include in vitro assays, such as cytotoxicity testing utilizing acceptable cell lines (e.g., liver cells) to evaluate direct cellular damage, and genotoxicity and mutagenicity testing that may not only establish DNA damage, but severity of any mutations (Chrz et al. 2020). For in vivo animal studies, the initial acute toxicity studies may be a single-dose study that allows calculation of an LD50 for the product; however, subacute and chronic toxicity studies should be performed as well to determine effects from repeated exposures over longer exposures while measuring the function of primary organs (e.g., liver function, kidney function, etc.), hematology, and histopathological effects on organs (Erhirhie et al. 2018). In addition to acute and chronic toxicity studies, development of products that are dermal applications (such as dyes or ingredients in cosmetics), dermal toxicity studies are also warranted to rule out skin irritation or systemic toxicity caused by absorption through the skin (Strickland et al. 2023). Moreover, for industrial development and capacity, componentry like effluent releases from processes like pulping or dyeing would require a comprehensive environmental assessment with respect to toxic constituents (Haile et al. 2021).

Standardization and quality control are mandatory aspects of safe product development. This includes the creation of detailed standard operating procedures (SOPs) for all steps from raw biomass harvesting, to drying, extraction, purification, and product formulation (Walker et al. 2018). The development of thorough chemical fingerprinting, such as through mass spectrometry or chromatography, aids in the development of consistent chemical profiles, as well as managing the undesirable levels of impurities and/or toxins in the final product. All herbal products, food additives, or industrial chemicals should comply with regulatory protocols (national and international) adopting industry-established Good Manufacturing Practices (GMP) and Good Laboratory Practices (GLP) (Donno et al. 2016). Finally, a total benefit/risk analysis should be conducted on all products. For medicinal applications, the potential therapeutic benefits must be demonstrably greater than any known potential or residual toxic effects. For industrial products, there must be some environmental and/or economically benefits to showing the management of any inherent toxicological hazards (Kürzinger et al. 2020).

11.4 Detoxification Strategies for *Lantana camara* Poisoning

Detoxification strategies are of great importance to treat animals and humans when they have been poisoned with *Lantana camara* (Kumar et al. 2016).

11.4.1 For Livestock Poisoning

Treatment is mainly concerned with a lot of supportive care and removing the reasons for the continued poisoning of their animals. The most important thing to do is immediately remove the animal from areas infested with *Lantana camara* (Varga and Puschner 2012). Supportive therapy can also consist of providing intravenous or oral fluids to avoid dehydration, where dehydration is often caused by the combination of anorexia and constipation. Gastrointestinal management may cover the use of laxatives (i.e., magnesium sulfate) (Baruti et al. 2018). These treatments may also accelerate the speed of passage of their ingesta, including unabsorbed, poison-containing ingesta to the lower gut. Should a ruminant experience a significant rumen impaction, the answer may lie in a rumenotomy (surgical removal of rumen contents). The injection of hepatoprotectants (e.g., methionine, choline, thioctic acid, the herbal hepato-protectant silymarin) to hasten liver regeneration may prove useful, although there is often limited success for very severe cases. For photosensitization, it is important to keep affected animals in shaded areas to mitigate the chances of further UV light exposure. Topical sunscreens or barrier coverings (e.g., Charmil) can be placed on unpigmented skin lesions, and any secondary bacterial infection of the skin can be treated with an appropriate antibiotic cover. Nutritional support should include palatable and digestible high-quality feed. In cases of plant toxin ingestion in the early stages of poisoning, activated charcoal can be given orally in order to adsorb the toxins in the gastrointestinal tract and to limit their systemic absorption (Baruti et al. 2018). The prognosis for poisoned livestock depends solely on the toxicity of the poison and time to treatment; usually, animals with major hepatotoxicity or extensive photosensitisation result in guarded to poor prognosis. Ultimately, effective management of *Lantana camara* infected lands and adequate alternative forage remains the best option for farmers operating in affected areas (Baruti et al. 2018).

11.4.2   *For Human Poisoning (Accidental Ingestion)*

When a human (especially a child) accidentally ingests *Lantana camara* plants or plant parts, it requires emergency medical attention. Depending on the timing and the unintentional ingestion, there are emergency treatment steps that may need to be taken under the supervision of a medical professional. These emergency treatment steps may include gastric lavage (if the poisoning is very recent—usually within 1–2 hours) to remove any plant material that may not yet be absorbed from the stomach. In some cases, activated charcoal may be given orally to absorb toxins that are still in the gastrointestinal tract and as a poison preventative for systemic absorption (Sharma et al. 2007). Supportive care is very important and must include appropriate management of fluids and electrolytes so that dehydration and electrolyte changes from vomiting and diarrhea can be properly managed. In some cases, oral antiemetics may be provided to manage persistent nausea and vomiting. It remains very important to follow liver health and damage through regular blood work to obtain liver enzyme levels and bilirubin concentrations. Symptomatic care resulting from impacts on neurological, cardiac, or respiratory health must be monitored and treated. Most human cases of *Lantana camara* poisoning are mild and resolve with supportive care, meaning that while there is a poison event, it does not compromise health, and care provided is supportive in nature (Pingale 2020). Severe poisoning leading to liver failure or significant neurological issues is rare, but it does happen, it may need intensive care, and it may be fatal.

11.4.3   Detoxification of **Lantana camara** *Biomass for Value-Added Products*

In addition to treating acute poisoning, there are also developments to try to process raw *Lantana camara* biomass such that it has reduced or no toxic lantadene concentrations, so the products can be considered safe for intended use. In terms of physical processing, actual washing could remove surface contaminants, but contaminants that are internalized are likely not removed. Drying or applying heat treatment would degrade some compounds, but lantadenes are resistant to heat. Thus, this may not be enough to detoxify the biomass, though extreme heat, such as pyrolysis for biochar production, could destroy less resistant compounds (Barman and Roy 2022).

Alkaline treatments are chemical methods. Many industrial processes, such as the Kraft or soda pulping methods used to make paper, use strong alkaline conditions (Iglesias et al. 2020; Bhodiwal et al. 2024). This could potentially degrade, alter, or convert the lantadenes into less toxic or inactive form or could make it insoluble in the product. There is continuing work to optimize alkaline conditions for detoxification. Another form of chemical method is adsorption; activated charcoal or other adsorbent materials could be used in the extraction or processing stages to selectively bind and remove toxins from the product stream (Bhodiwal et al. 2022).

Biological methods, often under the realms of bioremediation or fermentation, are also being examined (Pandey et al. 2016). Specifically, research is focused on specific bacteria or fungi that can metabolize and degrade lantadenes. If successful, this subsequently untreated can serve as a biological pretreatment step for biomass that is intended for valorization or for treating toxic waste streams. Additionally, a regulated composting process can also degrade many organic toxins over time and produce thoroughly composted Lantana biomass, and using the microbial activity will allow for use as an organic fertilizer or as an organic soil amendment (Rawat and Suthar 2014). This would allow the lantadenes to decompose in the composting process. Also, as outlined in the previous chapter, selective extraction and fractionation are equally important. This means using solvents of different polarities and advanced chromatographic separation techniques during the extraction to concentrate beneficial compounds into one fraction and leave the toxic lantadenes in another discardable fraction. This method has a strong potential to develop safe medicinal extracts.

11.5   Conclusion

The toxicology of *Lantana camara* has been a key consideration in valuing this plant for value-added products. The inherent hepatotoxicity and photosensitization of livestock experienced, which is largely due to lantadenes, are common causes of the economic losses experienced in affected regions. Human poisoning events are rarer; however, it can and do occur, especially with unripe berries being ingested accidentally by children, and must be recognized as a serious possibility that requires attention.

Importantly, any proposal for using *Lantana camara* as a biomass source will need to be supported by a proactive and thorough risk assessment. Any such risk assessment will require systematic chemical characterisation, a determining quantitative value for toxins, in vitro and in vivo toxicity testing of all products, and good quality control. It will also require the ongoing research and development of detoxification products (either to treat poisoned animals or to detoxify plant biomass for industrial/product use). While *Lantana camara* offers an opportunity, the potential for sustainable use is entirely dependent on knowledge and management of its toxicology to ensure benefits while protecting human, animal, and environmental health.

References

Barman KK, Roy D. Utilization of weed biomass. Technological glimpses on weeds and their management. In: 3rd international weed conference. Anand: AAU; 2022. p. 58–60.

Baruti M, Borthakur A, Chutia J, Singh B, Bhuyan M, Bhuyan D. Management of *Lantana camara* poisoning in a bull. Int J Chem Sci. 2018;6(1):950–2.

Bhodiwal S, Agarwal R, Chauhan S. Evaluation for characterization of pulp and paper making from weed species *Lantana camara* L. Biospectra. 2022;17(2):187–92.

Bhodiwal S, Agarwal R, Chauhan S. A weed species Lantana camara L.: an alternative raw material in handmade paper making. TWIST. 2024;19(1):268–74.

Carstairs SD, Luk JY, Tomaszewski CA, Cantrell FL. Ingestion of *Lantana camara* is not associated with significant effects in children. Pediatrics. 2010;126(6):me1585–8. https://doi.org/10.1542/peds.2010-1669.

Chrz J, Hošíková B, Svobodová L, Očadlíková D, Kolářová H, Dvořáková M, Kejlová K, Malina L, Jírová G, Vlková A, Mannerström M. Comparison of methods used for evaluation of mutagenicity/genotoxicity of model chemicals – parabens. Physiol Res. 2020;69(Suppl 4):S661–79. https://doi.org/10.33549/physiolres.934615.

Day, M. D., Wiley, C. J., Playford, J., & Zalucki, M. P. Will biological control of Lantana camara ever succeed? Patterns, processes & prospects. Biological Control. 2003;28(3), 291–322. https://doi.org/10.1016/S1049-9644(03)00104-5.

Donno D, Boggia R, Zunin P, Cerutti AK, Guido M, Mellano MG, Prgomet Z, Beccaro GL. Phytochemical fingerprint and chemometrics for natural food preparation pattern recognition: an innovative technique in food supplement quality control. J Food Sci Technol. 2016;53(2):1071–83. https://doi.org/10.1007/s13197-015-2115-6.

Erhirhie EO, Ihekwereme CP, Ilodigwe EE. Advances in acute toxicity testing: strengths, weaknesses and regulatory acceptance. Interdiscip Toxicol. 2018;11(1):5–12. https://doi.org/10.2478/intox-2018-0001.

Haile A, Gelebo GG, Tesfaye T, Wassie M, Mebrate MA, Abuhay A, Limeneh DY. Pulp and paper mill wastes: utilizations and prospects for high value-added biomaterials. Bioresour Bioprocess. 2021;8:35. https://doi.org/10.1186/s40643-021-00385-3.

Iglesias MC, Gomez-Maldonado D, Via BK, Jiang Z, Peresin MS. Pulping processes and their effects on cellulose fibers and nanofibrillated cellulose properties: a review. For Prod J. 2020;70(1):10–21.

Kato-Noguchi H, Kato M. Compounds involved in the invasive characteristics of *Lantana camara*. Molecules. 2025;30(2):411. https://doi.org/10.3390/molecules30020411.

Kumar R, Katiyar R, Kumar S, Kumar T, Singh V. *Lantana camara*: an alien weed, its impact on animal health and strategies to control. J Exp Biol Agric Sci. 2016;4:321–37. https://doi.org/1 0.18006/2016.4(3S).321.337.

Kumar R, Sharma R, Patil RD, Mal G, Kumar A, Patial V, Kumar P, Singh B. Sub-chronic toxico-pathological study of lantadenes of Lantana camara weed in Guinea pigs. BMC Vet Res. 2018;14(1):129. https://doi.org/10.1186/s12917-018-1444-x.

Kürzinger ML, Douarin L, Uzun I, El-Haddad C, Hurst W, Juhaeri J, Tcherny-Lessenot S. Structured benefit-risk evaluation for medicinal products: review of quantitative benefit-risk assessment findings in the literature. Ther Adv Drug Safety. 2020;11:2042098620976951. https://doi.org/10.1177/2042098620976951.

Negi GCS, Sharma S, Vishvakarma SCR, Samant SS, Maikhuri RK, Prasad RC, Palni LMS. Ecology and use of *Lantana camara* in India. Bot Rev. 2019;85:109–30. https://doi. org/10.1007/s12229-019-09209-8.

Ntalo M, Ravhuhali KE, Moyo B, Hawu O, Msiza NH. *Lantana camara*: poisonous species and a potential browse species for goats in southern Africa – a review. Sustainability. 2022;14(2):751. https://doi.org/10.3390/su14020751.

Pandey SK, Bhattacharya T, Chakraborty S. Metal phytoremediation potential of naturally growing plants on fly ash dumpsite of Patratu thermal power station, Jharkhand, India. Int J Phytoremediation. 2016;18(1):87–93.

Pass MA. Current ideas on the pathophysiology and treatment of *Lantana* poisoning of ruminants. Aust Vet J. 1986;63:169–71.

Pingale SS. Toxicity study of *Lantana camara* leaves. Pharm Lett. 2020;13(2):22–9.

Pour BM, Latha LY, Sasidharan S. Cytotoxicity and oral acute toxicity studies of Lantana camara leaf extract. Molecules. 2011;16(5):3663–74. https://doi.org/10.3390/molecules16053663.

Rawat I, Suthar S. Composting of tropical toxic weed *Lantana camera* L. biomass and its suitability for agronomic applications. Compost Sci Util. 2014;22:105–15. https://doi.org/10.108 0/1065657X.2014.895455.

Rowe LD. Photosensitization problems in livestock. Vet Clin N Am Food Anim Pract. 1989;5:301–23.

Sharma OP, Makkar HPS. Lantana-the foremost livestock killer in Kangra district of Himachal Pradesh. Livestock Adviser. 1981;6:29–31.

Sharma OP, Sharma PD. Natural products of the lantana plant-the present and prospects; 1989.

Sharma OP, Makkar HP, Dawra RK, Negi SS. A review of the toxicity of *Lantana camara* (Linn) in animals. Clin Toxicol. 1981;18(9):1077–94. https://doi.org/10.3109/15563658108990337.

Sharma OP, Sharma S, Pattabhi V, Mahato SB, Sharma PD. A review of the hepatotoxic plant *Lantana camara*. Crit Rev Toxicol. 2007;37(4):313–52. https://doi.org/10.1080/10408440601177863.

Shyamkumar TS, Aneesha VA, Kesavan M, Kumar D. *Lantana*: the dark story behind an alluring plant. epashupalan. 2021;3(2):95–8. https://wp.me/pbYZMt-2tk

Stegelmeier BL, Field R, Panter KE, Hall JO, Welch KD, Pfister JA, Gardner DR, Lee ST, Colegate S, Davis TZ, Green BT, Cook D. Selected poisonous plants affecting animal and human health. In: Wanda MH, Colin GR, Matthew AW, editors. Haschek, Rousseaux handbook of toxicologic pathology. 3rd ed. Academic Press; 2013. p. 1259–314. https://doi.org/10.1016/B978-0 -12-415759-0.00040-6.

Strickland J, Haugabrooks E, Allen DG, Balottin LB, Hirabayashi Y, Kleinstreuer NC, et al. International regulatory uses of acute systemic toxicity data and integration of new approach methodologies. Crit Rev Toxicol. 2023;53(7):385–411. https://doi.org/10.1080/10408444.202 3.2240852.

Varga A, Puschner B. Retrospective study of cattle poisonings in California: recognition, diagnosis, and treatment. Vet Med Res Rep. 2012;3:111–27. https://doi.org/10.2147/VMRR.S28770.

Walker SM, Pearson TRH, Casarim FM, Harris N, Petrova S, Grais A, Swails E, Netzer M, Goslee KM, Brown S, Sidman G. Standard operating procedures for terrestrial carbon measurement. Version 2018. Winrock International; 2018.

Chapter 12
Balancing Control and Management of *Lantana camara*: Integrative Approaches and Future Directions

Abstract This chapter gives an extensive overview of current control and management techniques for *Lantana camara*, including mechanical, chemical, biological, and the important role of integrative methods. We describe the effectiveness, limitations, and recent developments in each method and ethical concerns that arise when trying to manage extensive invasion. Socioeconomic aspects of *Lantana camara* invasion are also covered, and we describe some use-based management strategies. The overall message is that successful long-term management of *Lantana camara* is through an integrated, adaptive, and community-based framework that combines short-term suppression, ecological restoration, and sustainable utilization.

12.1 Introduction

Lantana camara, a perennial shrub from the American tropics, is among the 100 worst invasive alien species in the world. Introduced all over the world for many purposes, especially as an ornamental plant and to control erosion, its presence is now recognized as an ecological catastrophe. Free to invade vast areas in the tropics and subtropics, it has spread into forests, grasslands, croplands, urban areas, and degraded lands (Bogale and Tolossa 2021).

The effects of invasion from *Lantana camara* are complex and multiple—it ecological impacts are obvious; it produces dense structures which shade out and displace native vegetation, reduced biodiversity, and thereby altered ecosystem structure and performance, as well as threatening or displacing key habitat, decreasing forage for wild herbivores and livestock, and potentially mobilizing or blocking the movement of wildlife (De Lacy and Shackleton 2017). Economically, its impact leads to diminished agricultural productivity (e.g., variable levels of crop loss); a diminishment in the availability of non-timber forest products (NTFPs) is critical for sustaining local livelihoods, as well as any direct costs from NTFP use, management, and control. The plant is also known to be toxic to nonhuman animals, resulting in poisoning upon ingestion, primarily of its berries and leaves. In addition to this, *Lantana camara* can alter nutrient cycling in soils in some areas and increase

M. A. Dervash et al., *Lantana Camara*, SpringerBriefs in Plant Science,
https://doi.org/10.1007/978-3-032-03837-1_12

fire hazard at times by adding much fuel load, as well as create logistical problems for forest patrolling and management. Given these consequential implications, controlling and managing *Lantana camara* effectively is an important implication for environmental protection and sustainable development in the areas, if not the entire globe (Abebe 2018). This chapter will provide a much-needed synthesis and assessment of the current methods and actions implemented to control a formidable invasive weed.

12.2   Characteristics Contributing to Invasiveness

The biological and ecological characteristics of *Lantana camara* that facilitate its success as an invasive species are:

- *Rapid Growth Rate and Prolific Reproduction: Lantana camara* demonstrates exceptional growth rates, and growth into the canopy occurs relatively quickly. *Lantana* has a long flowering period, generally producing numerous small, berry-like drupes. Each fruit is commonly formed by a single ovary and usually contains two seeds. A single mature *Lantana camara* can produce many thousands of seeds per annum (Negi et al. 2019).
- *Efficient Seed Dispersal:* The fleshy fruits of *Lantana camara* are attractive to frugivorous birds and other herbivorous animals (Day et al. 2003a, b). The birds eat the fruits and deposit viable seeds in their faecal droppings after scarification of dormant endocarp of *Lantana camara* fruit, thus supporting Lantana's colonizing capabilities into new, often untouched areas, including remote areas of forest interiors (Buckley et al. 2006).
- *Coppicing Ability: Lantana camara* possesses severe coppicing ability, allowing it to return from the roots and base of stems whether it is cut or burnt. *Lantana camara* usually has a precutting or burning growth period which can correspond to the extensive destruction of *Lantana camara* in specific locations through traditional mechanical methods (Love et al. 2009).
- *Allelopathy:* The plant releases allelochemical compounds (e.g., phenolic acids, alkaloids) into the soil, or through leachate from the leaves, inhibiting germination and growth of neighboring species, giving *Lantana camara* a competitive advantage through suppression of native flora (Kato-Noguchi and Kurniadie 2021).
- *High Adaptability and Resilience: Lantana camara* is adaptable to a wide range of environmental conditions; it grows in a variety of habitats, including open spaces that are disturbed, edges of forests, agricultural lands, and degraded sites. *Lantana camara* is considered resilient, as it tolerates disturbance and unfavorable conditions. For example, it has resilience to droughts, fires (it typically regenerates with considerable vigour after fires), and nutrient-poor soils (Kohli et al. 2006). One of the few natural constraints on the spread of *Lantana camara* in temperate climates is frost and very cool temperatures.

- *Lack of Natural Enemies:* The natural predators, herbivores, and pathogens that check *Lantana camara* populations in its native habitat are often absent in its introduced range. This "enemy release" is a major driver of the serious expansion of invasive species (Kumar and Singh 2020).

These combined traits create a formidable invasive species, making its control a complex and ongoing challenge. All of the above characteristics of *Lantana camara* make a management approach for invasive species a complex, continually adapting problem.

12.3 Control and Management Strategies

The complex nature of the invasion by *Lantana camara* is best handled with multiple control and management options. No single treatment has been shown to work in all scenarios, with efficacy being best achieved with an integrated approach to management, which is site- and infestation-specific (Negi et al. 2019; Bhagwat et al. 2012).

12.3.1 *Mechanical/Manual Control*

Mechanical and manual control involves physically removing or damaging the *Lantana camara* plants. These actions are typically useful in smaller infestations, or in sensitive sites where herbicides could not be used, or for assistance to control agents of a more comprehensive management plan.

12.3.1.1 Hand Pulling

Hand pulling is most effective to do when the soil is moist; however, once the entire root system has been removed, it prevents the chance for coppicing. This method is very labor-intensive and unrealistic for mature plants or large areas that are infested (Love et al. 2009).

12.3.1.2 Slashing/Chopping

Slashing/chopping *Lantana camara* is simply cutting the stem, leaving the *Lantana camara* plant above ground. While this may seem simple, this method is usually ineffective because *Lantana camara* can regrow from dormant buds and will tend to grow denser and problematic thickets after this method is practiced. Additionally, slashing/chopping can stimulate the soil seed bank, which may release previously dormant seeds, resulting in more seedlings (Love et al. 2009).

12.3.1.3   Manual Grubbing

This is the use of digging to lift out the entire root system of *Lantana camara* clumps and is better than slashing (Bartholomew and Armstrong 1978). This is a labor-intensive activity and causes a high degree of soil disturbance. It can also expose buried *Lantana camara* seeds to light and increase the likelihood of mass germination. Manual grubbing also increases the chance of soil erosion, particularly on sloping sites, and damages any native species established nearby between clumps (Love et al. 2009).

12.3.1.4   Cut Rootstock Method (CRS)

A more sophisticated form of mechanical control that shows promise. This method cuts the main taproot of the *Lantana camara* plant only 3–5 cm below the soil surface, where the plant can regenerate, and below the coppicing zone, which is the transition point between the stem base of *Lantana camara* to the rootstock (Love et al. 2009). This will allow the clump to be removed, turned upside down, and dried. Cutting the taproot aims to mitigate the likelihood of regeneration from the rootstock, is less destructive to the surrounding area than manual grubbing, and results in less soil disturbance. Some positives of CRS are that it is cheaper, more straightforward than grubbing, does not utilize herbicides or exotic biological controls, and avoids deep soil disturbance (Love et al. 2009).

12.3.2   Limitations of Mechanical Control

The main limitations are high labor requirements, meaning costs prohibitive and impractical at larger infestations, the ground disturbance could instigate mass germination of the large seed bank and therefore require repeated follow-up control. Possibly the greatest challenge is regeneration from root fragments that were not removed completely or seeds that were disturbed during eradication.

12.3.3   Chemical Control

Chemical herbicides create a viable option for controlling *Lantana camara* in larger areas, where manual control will not be practical. Efficacy of herbicides against *Lantana camara* can vary owing to the herbicides themselves, the method of application, types of *Lantana camara*, and environmental conditions (Ferrell et al. 2012).

12.3.3.1   Herbicides and Efficacy

(i) *Glyphosate:* Marginally effective as a foliar application, often associated with regrowth.

(ii) *Fluroxypyr + Aminopyralid:* Good efficacy when applied twice in 6 months (likely expensive) but effective for fluroxypyr as a basal application (to the stem).

(iii) *Aminocyclopyrachlor:* Shows excellent control, with 100% control achieved with two applications—with effective control from one application.

12.3.3.2   Application Methods

(i) *Foliar Spray:* Application of the herbicide directly to the leaves. Foliar spraying occurs when the herbicide is sprayed to the point of runoff and is effective on actively growing plants that are under 2 meters tall (Tu et al. 2001).

(ii) *Basal Bark Spraying:* Application of a concentrated herbicide mixture to the bark surrounding the base of the stem. Basal bark spraying can be performed on single (one stem) or multistemmed plants (many stems) at any time of the year, though it is best performed in the fall when the target plant is actively growing. Multistemmed plants require application to the base of each stem (Tu et al. 2001).

(iii) *Cut-Stump Method:* The application of herbicide directly to the freshly cut surface of the stem. This application is used to prevent the regrowth of the stump when cut (Tu et al. 2001).

(iv) *Splatter Gun Techniques:* Applicators may find these techniques more effective for obtaining hard-to-reach places (Tu et al. 2001).

12.3.3.3   Drawbacks of Chemical Control

While chemical methods can be effective, there are various limitations:

(i) *Environmental Risks*: Risk to native biota, nontarget plants and water.

(ii) *Cost-Effectiveness:* Cannot be used for excessively large and infested sites as it is impractical and expensive.

(iii) *Herbicide Resistance:* Regular use of the same herbicide type may allow the population of *Lantana camara* to become herbicide resistant.

(iv) *Toxicity and Palatability:* Depending on the herbicide and its environmental stresses, it may be that chemicals may affect the *Lantana camara*, and this affects sugar in the leaves, making it more palatable, and there are grazing animals who may increase the risk of toxicity.

(v) *Application Inconsistencies:* Results may be difficult to ascertain due to variable applications in methods, application mix rates, or seasonality.

12.3.4 Biological Control

Biological control is the introduction of natural enemies (insects, pathogens) to the invaded environment from the invasive organism's native range for the purpose of population suppression. In biological control, the aim is not to eradicate, the aim is a long-term sustainable reduction of the weed's vigor and reproductive potential (Stenberg et al. 2021).

(i) *Principle:* In the native range of *Lantana camara*, there have been coevolved natural enemies that keep the *Lantana camara* populations in check. The idea with biological control is to introduce agents that are host specific to reestablish a level of natural control (facilitative biocontrol).

(ii) *Biocontrol Agents:* Worldwide, there have been more than 40 insect species and a few pathogens that have been released against *Lantana camara*. Major biocontrol agents include:

- *Insects: Teleonemia scrupulosa* (sap-sucking bug) (Muniappan et al. 1996), *Uroplata girardi* (Denton et al. 1991) and *Octotoma scabripennis* (leaf-mining beetles) (Day et al. 2003a), *Ophiomyia lantanae* (seed-feeding fly) (Vivian-Smith et al. 2006), *Leptobyrsa decora* (bug) (Harley 1971), and several moths. Other agents include *Calycomyza lantanae* (fly) (Julien and Griffiths 1998), *Lantanophaga pusillidactyla* (moth) (Denton et al. 1991), *Epinotia lantana* (moth) (Harley 1971), *Octotoma championi* (beetle), *Uroplata fulvopustulata* (beetle) (Day et al. 2003a, b), *Phenacoccus parvus* (bug), *Aconophora compressa* (bug) (Julien and Griffiths 1998), and *Falconia intermedia* (bug) (Day and Zalucki 2009a, b).
- *Pathogens:* Some research has been conducted using pathogens such as *Rhizopus* spp.—in combination with voracious feeder insects such as *Spilosoma obliqua*, it is possible to completely kill the weed (Tripathi et al. 2017).
- *Use of Goats:* Goats are resistant to *Lantana camara* poisoning; thus, it is recommended to use goats as a biological control in the infested areas (Ntalo et al. 2022).

(iii) *Effectiveness and Limitations:* Biocontrol of *Lantana camara* has had mixed success worldwide. In some instances, agents have successfully established and had localized effects, resulting in localized dieback or reduced seeding, but on the whole, sufficient numbers for agents to exert widespread control have not been achieved (Day and Zalucki 2009a, b). The biological control of *Lantana camara* has encountered the following challenges:

- *Host specificity, mismatch in native and introduced ranges:* Important precursor to a successful introduction of biocontrol agents is to be certain that they are entirely specific to *Lantana camara* and that they do not pose a risk to either native plant species or economically important plants. Day and Neser (2000) identify that if biocontrol agents were obtained from one of the native range *Lantana camara* species (e.g. Mexico, Caribbean), these agents could be less likely to adapt to hybrid populations (or variants) in the introduced region.

- *Varietal diversity or hybrid swarms: Lantana camara* has a lot of genetic diversity and has many varieties, and some biocontrol agents are only specific to one or two varieties, which limits the effectiveness of their use against all *Lantana camara* varieties. In many countries where *Lantana camara* occurs, it is represented as hybrid swarms that make it impossible to use biocontrol agents, because of the inability to match a consistent ontology (phenotype) of the host (Baars and Heystek 2003).
- *Environmental factors:* The efficacy of biocontrol agents may be highly dependent upon climatic factors (temperature, humidity, precipitation) ideal in the invaded area, which may vary considerably from their native areas.
- *Slow impact, agent specificity vs adaptability:* Biological control can be a strategy that takes years to decades to yield significant impacts. The success of oligophagous agents can be established for long periods with the host *Lantana camara*, but long-term effectiveness and implications for nontarget species will need continual evaluation (Paynter 2024).

12.3.5 *Modern Research Directions*

In regard to the issues presented above, the research and management now encompass:

 (i) *Molecular Taxonomy*: Two established techniques (DNA barcoding, genome-wide association studies (GWAS) and next-generation sequencing) are aiming to further reduce species boundaries and hybrid delineation (Bidyananda et al. 2024).
 (ii) *Integrated Pest Management* (IPM): Appropriate biological control agents would be used, combined with mechanical and chemical methods to limit damage further on various landscapes (Zhou et al. 2024).
 (iii) *International Collaborations*: By utilizing shared databases and potentially conducting joint fieldwork at sites of *Lantana camara*-infested countries, the biocontrol agents that would act globally at controlling *Lantana camara* effectively can be selected (Urban et al. 2011).

12.3.6 *Policy and Ecological Implications*

Lantana camara's ecological dominance and disturbance-resilience indicate that inaction with regard to its taxonomic complexity could have significant ramifications for control programs globally (Shackleton et al. 2017):

(i) *Ecosystem Disruption*: *Lantana camara* has the ability to replace the native flora, opportunistically engage or restrict fire regimes, and reduce biodiversity that will all require control to prioritize conservation (Shackleton et al. 2017).

(ii) *Resource Allocation*: Misidentifying *Lantana* forms could incur costs in resources with regard to selecting multiple excess control agents, every trial and release, and finally monitoring (Shackleton et al. 2017).

12.3.7   *Integrated Management Strategies (IMS)*

Since all individual control methods have limitations, an integrated management strategy (IMS) is generally viewed as the most effective and sustainable form of control for *Lantana camara* (Nanjappa et al. 2005). IMS uses combinations of different methods to utilize their synergies, target all growth stages, and deal with the specific features of the invasion.

(i) *Core Principles:* IMS for *Lantana camara* typically involves:

- *Initial Suppression:* Initially, a combination of mechanical (cut rootstock approach) and chemical approaches is sometimes used to achieve initial suppression of dense *Lantana camara* thickets, which reduces the biomass immediately and opens up the area (Love et al. 2009; Ferrell et al. 2012).
- *Follow-Up Treatment:* Follow-up treatments, where the area is managed after initial clearing, are important for managing regrowth from rootstock and new seedlings emerging from the soil seed bank. Follow-up treatment can be spot-spraying herbicides, hand-pulling young plants, or repeated cutting (Nanjappa et al. 2005).
- *Biological Control Integration:* Where possible, releasing a biocontrol agent that has been established will give long-term self-sustaining pressure on *Lantana camara* and should decrease the vigor of the weed and reproductive output over the long term, especially in areas too big for mechanical or chemical approaches (Stenberg et al. 2021).
- *Ecological Restoration:* A valuable and often karst aspect of an IMS. It is vitally important to make active restoration efforts: e.g., planting native species, reestablishing pastures and grazing, and improving soil health to prevent reinvasion by *Lantana camara* or secondary invasive species. A native ecosystem will provide competition and limit the potential niches for the weed (Negi et al. 2019).
- *Fire Management:* Controlled burning may feel more controversial compared to other methods but could be incorporated into an IMS, especially in fire-prone ecosystems. Fire can reduce existing *Lantana camara* biomass and potentially germinate seeds in the soil seedbank to allow better mechanical or chemical treatments of new seedlings (Debuse and Lewis 2014). A burning will have to be orchestrated well, as it can have unwanted consequences on native biodiversity and the soil.

- *Prevention and Monitoring:* Prevention and timely and rapid response, i.e., early detection, follow-up, and treatment of new infestation, are the best and most cost-effective methods. Continuous surveillance and monitoring of removed areas allow for tracking recovery responses, detection of new outbreaks, and adaptive management (Gooden et al. 2009).

IMS understands that a fix is never a fix and continuing effort, plus the need for flexibility and combinations of approaches to achieve effective and lasting *Lantana camara* management.

12.4 Challenges in *Lantana camara* Management

Despite ongoing investigation and global management effort, *Lantana camara* remains an exceptionally difficult invasive species to manage. There are various reasons, comprising of both plant characteristics and external factors, that contribute to the inability to manage Lantana effectively, such as, high reproductive capacity, persistent soil seed bank, vigorous coppicing, allelopathic effects, scale of infestation, environmental adaptability, ecosystem alteration, economic and resource constraints lack of public awareness and participation, monitoring and follow-up fatigue. These challenges reflect the need for new, lasting, and integrated solutions requiring significant ecologically restorative inputs and community engagement.

12.5 Socioeconomic Aspects and Utilization

The invasion of *Lantana camara* has considerable socioeconomic issues, especially for rural and forest-dependent communities. Emerging speculation also examines possibilities for use as a method for offsetting the costs of control and stimulating removal.

12.5.1 Negative Socioeconomic Impacts

12.5.1.1 Loss to Agriculture

Lantana camara invades agricultural lands and competes with crops for inputs. It lowers economic yield and therefore incurs a loss directly to the farmer (Hamad et al. 2022).

12.5.1.2   Loss to Nontimber Forest Products (NTFPs)

Forest-dwelling communities depend on NTFPs (i.e., medicinal plants, fruits, fodder) for their livelihoods, and *Lantana camara* invades and displaces native NTFP-yielding species. This can be a sea change for local economies as the NTFP potential diminishes significantly. One example, studies in central India demonstrate statistically significant negative correlations of the plant and distance from *Lantana camara* invaded plots measuring crop loss and evidence that demonstrates significant loss in yields for community members collecting NTFP (Kannan et al. 2016).

12.5.1.3   Livestock Poisoning

The plant is poisonous to livestock and can cause significant health issues such as liver failure, causing mortality, which leads to substantial negative economic losses to pastoralists (Sharma et al. 2007).

12.5.1.4   Impact on Human Health and Livelihoods

Removal with hand tools to manually eliminate *Lantana camara* is very physical, and workers are at risk of exposure to thorns and skin irritations, which may exacerbate their low-paid socioeconomic status (Kumar and Singh 2020). *Lantana camara* infestation may also make forest patrolling and management more difficult.

12.5.2   Potential for Utilization-Based Management

Understanding that control incurs high costs means that research is being done to find economic uses for *Lantana camara* biomass. This "weed to wealth" approach seeks to create value-added products that provide incentives for widespread removal and shift *Lantana camara*'s perception from a liability to a resource (Kumar and Singh 2020).

12.5.2.1   Biofuel and Briquettes

The woody biomass of *Lantana camara* has the potential for conversion to briquettes or for raw material to produce biofuels, providing an alternative energy source (Koricho et al. 2017).

12.5.2.2 Handicrafts and Furniture

Lantana camara's durable stems have been used in some areas to produce baskets or furniture and utility articles. Developing such initiatives at a larger scale could generate livelihood for local communities (Priyanka et al. 2013).

12.5.2.3 Paper Making

Preliminary literature and studies suggest that *Lantana camara* fibers have ideal strength properties to serve as raw material for making paper, including specialty papers. With relatively low cost and local handmade paper industries, *Lantana camara* can present good income generation opportunities (Bhodiwal et al. 2024).

12.5.2.4 Biochar and Soil Amendment

Lantana camara biomass can be designated to produce biochar via pyrolysis, and biochar can act as a soil amendment material by enhancing soil fertility and carbon sequestration (Shyam et al. 2025).

12.5.2.5 Compost

Compostable (perhaps producing compost with allelopathic properties), but degrading the local species of *Lantana camara* biomass via true compost standards can provide valuable organic fertilizer (Rawat and Suthar 2014).

The practice of the utilization strategy offers a win-win situation by managing the invasive species while also generating economic returns for impacted communities. However, careful consideration must be given to the sustainability of these initiatives, ensuring that they do not inadvertently foster the growing and spreading of *Lantana camara* and that there is demand for the products the different jobs offer.

12.6 Conclusion

Lantana camara remains one of the most widespread invasive species with continuing and increasing threats to biodiversity, ecosystem services, and livelihoods across its range. This review of modern controls and management methods shows that there is no single solution to these invasive *Lantana camara* species. Control methods that are mechanical, chemical, and biological have their own pros and cons. The success of these methods also depends on the area of *Lantana camara* infestation (large or small), where it is located, and what resources are available to do it.

Best practice of *Lantana camara* management in contemporary times focuses on Integrated Management Strategies (IMS). As an overall strategy, initial knockdown (mechanically using cut rootstock or chemical options) is essential. This needs to be followed up with continued follow-up, potentially integrating biological controls for sustained pressure and the all-important ecological restoration. Restoration actions include reestablishing native vegetation and/or ecological processes. This is important to limit reinvasion from the resistant soil seed bank and dispersed seeds around the landscape.

Moreover, *Lantana camara* has serious socioeconomic cost implications, and therefore, community involvement and the possible approach of management based on use are essential. Community involvement can develop productive actions with *Lantana camara*. Development of economically viable products from *Lantana camara* biomass (e.g., handicrafts, paper, biofuels) can help develop incentives for the removal process and hopefully turn a nasty weed into a resource for local sustainable livelihoods.

In the future, the ongoing research to identify more potent, host-specific biological control agents, continue to develop chemical formulations to reduce environmental toxicity and cost reduction, and improve mechanical methods for more effectiveness will be paramount. It is critical that researchers, policymakers, community members, and resource managers work together to create and implement adaptive, site-specific, and integrated management plans. While sustained management options are essential, long-term success in the management of *Lantana camara* will depend on sustained commitment to integrated strategies to not only control the invader but also to restore ecological integrity and provide agency among those affected.

References

Abebe FB. Invasive *Lantana camara* L. shrub in Ethiopia: ecology, threat, and suggested management strategies. J Agric Sci. 2018;10(7):184–95. https://doi.org/10.5539/jas.v10n7p184.

Bartholomew BL, Armstrong TR. A new look at Lantana control. Queensland Agricultural Journal. 1978;104(4), 339–344.

Baars JR, Heystek F. Biocontrol agents established on *Lantana camara* in South Africa. BioControl. 2003;48:743–59. https://doi.org/10.1023/A:1026389207517.

Bhagwat SA, Breman E, Thekaekara T, Thornton TF, Willis KJ. A battle lost? Report on two centuries of invasion and management of *Lantana camara* L. in Australia, India and South Africa. PLoS One. 2012;7(3):e32407.

Bhodiwal S, Agarwal R, Chauhan S. A weed species *Lantana camara* L.: an alternative raw material in handmade paper making. TWIST. 2024;19:268–74.

Bidyananda N, Jamir I, Nowakowska K, Varte V, Vendrame WA, Devi RS, Nongdam P. Plant genetic diversity studies: insights from DNA marker analyses. Int J Plant Biol. 2024;15(3):607–40. https://doi.org/10.3390/ijpb15030046.

Bogale GA, Tolossa TT. Climate change intensification impacts and challenges of invasive species and adaptation measures in eastern Ethiopia. Sustain Environ. 2021;7(1):1–24. https://doi.org/10.1080/23311843.2021.1875555.

Buckley YM, Anderson S, Catterall CP, Corlett RT, Engel T, Gosper CR, Nathan R, Richardson DM, Setter M, et al. Management of plant invasions mediated by frugivore interactions. J Appl Ecol. 2006;43:848–57. https://doi.org/10.1111/j.1365-2664.2006.01210.x.

Day M, Neser S. Factors influencing the biological control of *Lantana camara* in Australia and South Africa. In: Proceedings of the X international symposium on biological control of weeds. Bozeman: Montana State University; 2000. p. 897–908.

Day MD, Zalucki MP. *Lantana camara* Linn. (Verbenaceae). In: Muniappan R, Reddy GVP, Raman A, editors. Biological control of tropical weeds using arthropods. Cambridge University Press; 2009a. p. 211–46.

Day M, Zalucki M. *Lantana camara* Linn. (Verbenaceae): biological control of tropical weeds using arthropods; 2009b. https://doi.org/10.1017/CBO9780511576348.012.

Day MD, Broughton S, Hannan-Jones MA. Current distribution and status of *Lantana camara* and its biological control agents in Australia, with recommendations for further biocontrol introductions into other countries. Biocontrol News Inf. 2003a;24:63N–76N.

Day MD, Wiley CJ, Playford J, Zalucki MP. *Lantana*: current management status and future prospects. Canberra: Australian Centre for International Agricultural Research Canberra; 2003b.

De Lacy P, Shackleton C. Aesthetic and Spiritual Ecosystem Services Provided by Urban Sacred Sites. Sustainability, 2017;9(9):1628. https://doi.org/10.3390/su9091628.

Debuse V, Lewis T. Long-term repeated burning reduces *Lantana camara* regeneration in a dry eucalypt forest. Biol Invasions. 2014;16:2697–711. https://doi.org/10.1007/s10530-014-0697-y.

Denton GRW, Muniappan R, Marutani M. The distribution and biological control of *Lantana camara* in Micronesia. Micronesica Suppl. 1991;3:71–81.

Ferrell J, Sellers B, et al. Herbicidal control of Largeleaf Lantana (*Lantana camara*). Weed Technol. 2012;26:554–8. https://doi.org/10.2307/23264369.

Gooden B, French K, Turner P. Invasion and management of a woody plant, *Lantana camara* L., alters vegetation diversity within wet sclerophyll forest in southeastern Australia. For Ecol Manag. 2009;257(3):960–7. https://doi.org/10.1016/j.foreco.2008.10.040.

Hamad AA, Kashaigili JJ, Eckert S, Eschen R, Schaffner U, Mbwambo JR. Impact of invasive *Lantana camara* on maize and cassava growth in East Usambara, Tanzania. Plant Environ Interact. 2022;3(5):193–202. https://doi.org/10.1002/pei3.10090.

Harley KLS. Biological control of lantana. PANS. 1971;17:433–7.

Julien MH, Griffiths MW. Biological control of weeds: a world catalogue of agents and their target weeds. 4th ed. Wallingford: CABI Publishing; 1998. p. 223.

Kannan R, Shackleton C, Krishnan S, Shaanker R. Can local use assist in controlling invasive alien species in tropical forests? The case of *Lantana camara* in southern India. For Ecol Manag. 2016;376:166–73. https://doi.org/10.1016/j.foreco.2016.06.016.

Kato-Noguchi H, Kurniadie D. Allelopathy of *Lantana camara* as an invasive plant. Plants (Basel). 2021;10(5):1028. https://doi.org/10.3390/plants10051028.

Kohli R, Batish D, Singh H, Dogra K. Status, invasiveness and environmental threats of three tropical American invasive weeds (*Parthenium hysterophorus* L., *Ageratum conyzoides* L., *Lantana camara* L.) in India. Biol Invasions. 2006;8:1501–10. https://doi.org/10.1007/s10530-005-5842-1.

Koricho SA, Leta S, Soromessa T, Khan MM. Fuel briquette potential of *Lantana camara* L. weed species and its implications for weed management and recovery of renewable energy sources in Ethiopia. J Environ Sci Toxicol Food Technol. 2017;11:41–51.

Kumar RP, Singh JS. Invasive alien plant species: their impact on environment, ecosystem services and human health. Ecol Indic. 2020;111:106020. https://doi.org/10.1016/j.ecolind.2019.106020.

Love A, Babu S, Babu C. Management of *Lantana*, an invasive alien weed, in forest ecosystems of India. Curr Sci. 2009;97:1421–9.

Muniappan R, Denton GRW, Brown JW, Lali TS, Prasad U, Singh P. Effectiveness of the natural enemies of *Lantana camara* on Guam: a site and seasonal evaluation. Entomophaga. 1996;41:167–82.

Nanjappa HV, Saravanane P, Ramachandrappa BK. Biology and management of *Lantana camara* L.–a review. Agric Rev. 2005;26(4):272–80.

Negi GCS, Sharma S, Vishvakarma SCR, Samant SS, Maikhuri RK, Prasad RC, Palni LMS. Ecology and use of *Lantana camara* in India. Bot Rev. 2019;85:109–30. https://doi.org/10.1007/s12229-019-09209-8.

Ntalo M, Ravhuhali KE, Moyo B, Hawu O, Msiza NH. *Lantana camara*: poisonous species and a potential browse species for goats in southern Africa – a review. Sustainability. 2022;14(2):751. https://doi.org/10.3390/su14020751.

Paynter Q. Prioritizing candidate agents for the biological control of weeds. Biol Control. 2024;188:105396. https://doi.org/10.1016/j.biocontrol.2023.105396.

Priyanka N, Shiju MV, Joshi PK. A framework for management of *Lantana camara* in India. Proc Int Acad Ecol Environ Sci. 2013;3:306–23.

Rawat I, Suthar S. Composting of tropical toxic weed *Lantana camera* L. biomass and its suitability for agronomic applications. Compost Sci Util. 2014;22:105–15. https://doi.org/10.108 0/1065657X.2014.895455.

Shackleton RT, Witt A, Opio WA, Pratt CF. Distribution of the invasive alien weed, *Lantana camara*, and its ecological and livelihood impacts in eastern Africa. Afr J Range Forage Sci. 2017;34:1–11. https://doi.org/10.2989/10220119.2017.1301551.

Sharma OP, Sharma S, Pattabhi V, Mahato SB, Sharma PD. A review of the hepatotoxic plant *Lantana camara*. Crit Rev Toxicol. 2007;37(4):313–52. https://doi.org/10.1080/10408440601177863.

Shyam S, Ahmed S, Joshi SJ, Sarma H. Biochar as a soil amendment: implications for soil health, carbon sequestration, and climate resilience. Discover Soil. 2025;2:18. https://doi.org/10.1007/s44378-025-00041-8.

Stenberg JA, Sundh I, Becher PG, Björkman C, Dubey M, Egan PA, Friberg H, Gil JF, Jensen DF, Jonsson M, Karlsson M, Khalil S, Ninkovic V, Rehermann G, Vetukuri RR, Viketoft M. When is it biological control? A framework of definitions, mechanisms, and classifications. J Pest Sci. 2021;94:665–76. https://doi.org/10.1007/s10340-021-01354-7.

Tripathi M, Kumar V, Kulkarni N, Pauranik M. Biological control of *Lantana camara* through *Spilosoma oblique* and *Rhizopus* species. Asian J Exp Sci. 2017;31:27–30.

Tu M, Hurd C, Randall JM. Weed control methods handbook: tools and techniques for use in natural areas. Davis: Wildland Invasive Species Program, The Nature Conservancy; 2001. p. 195.

Urban AJ, Simelane DO, Retief E, Heystek F, Williams HE, Madire LG. The invasive 'Lantana camara L.' hybrid complex (Verbenaceae): a review of research into its identity and biological control in South Africa. Afr Entomol. 2011;19(2):315–48.

Vivian-Smith G, Gosper CR, Wilson A, Hoad K. *Lantana camara* and the fruit- and seed-damaging fly, *Ophiomyia lantanae* (Agromyzidae): seed predator, recruitment promoter or dispersal disruptor? Biol Control. 2006;36:247–57.

Zhou W, Arcot Y, Medina RF, Bernal J, Cisneros-Zevallos L, Akbulut MES. Integrated pest management: an update on the sustainability approach to crop protection. ACS Omega. 2024;9(40):41130–47. https://doi.org/10.1021/acsomega.4c06628.

Glossary

Agricultural pest is any organism that adversely affects cultivated plants or animals, adhering to agricultural practices that result in decreased yield or quality. Agricultural pests can include insects, weeds, fungi, rodents, and birds, which compete with or damage agricultural production.

Allelopathic properties are the ability of the plant to produce biochemicals (allelochemicals) that impact the growth, survival, or reproduction of organisms nearby. This effect could be inhibitory and may affect competing plants' growth, but it could also be stimulatory.

Analgesics are medications that alleviate pain while still leaving the user fully conscious of their surroundings. These medications exert their analgesic properties via numerous mechanisms that alleviate pain, from mild aches to severe pain.

Anti-inflammatories are substances that lessen inflammation. Inflammation is a natural response of the body to injury or infection that is accompanied by pain, swelling, heat, and redness. Anti-inflammatory agents exert their anti-inflammatory characteristics by blocking certain pathways along the inflammatory pathway.

Antipyretics are medications that reduce fever. Antipyretics work by lowering body temperature, usually through effects on the hypothalamus in the brain.

Bioactive compounds can be defined as a substance that may be present in living organisms and has a biological effect in another living organism. Bioactive compounds may confer health benefits and relate to pharmaceutical, medicinal/health, or nutritional applications.

Biodiversity is the diversity of life on Earth, including everything from genes to ecosystems, and the ecological processes that maintain it. It includes the diversity found within species, between species, and within ecosystems. In biological terms, everything is connected in the web of life that draws a distinction between life that is both very similar along an evolutionary lineage (causal) and life that is quite dissimilar (non-causal).

© The Editor(s) (if applicable) and The Author(s), under exclusive license to
Springer Nature Switzerland AG 2025
M. A. Dervash et al., *Lantana Camara*, SpringerBriefs in Plant Science,
https://doi.org/10.1007/978-3-032-03837-1

Bioherbicides are agents that are biologically based, often from microorganisms such as fungi or bacteria or plant extracts, to be used to control weeds in agriculture. They are environmentally friendly because they tend to target the unwanted plant better than chemical herbicides.

Biopesticides are materials (usually naturally occurring), substances, or organisms (e.g., certain bacteria, fungi, plant extracts, etc.) that can be used as pest controls for insects, weeds, and diseases. Biopesticides offer a more environmentally friendly technology than synthetic chemical methodologies. Biopesticides usually target a more specific pest and tend to persist in the ambient environment for shorter periods than synthetic chemicals.

Bioprospecting is the explorative research for useful natural products and genetic material derived from living organisms, such as microorganisms and plants, for further commercial purposes. Bioprospecting is used by industries such as pharmaceuticals, agriculture, and cosmetics to find new drugs, enzymes, and other useful products.

Carbon sequestration is the long-term removal, capture, or storage of carbon dioxide from the atmosphere, typically in geological formations, oceans, or biological systems such as forests and soils. The primary aim of carbon sequestration is to lessen climate change by reducing greenhouse gases in the atmosphere.

Decoction is a technique of extraction where the plant material (especially tough bits like roots or bark) is boiled in water to dissolve and concentrate its chemical components. The end result is a liquid (also called a decoction) that can be used in traditional medicine for its therapeutic qualities.

Ecological liability is the liability and or responsibility related to environmental harm or decay that resulted from an entity's actions, including cost of remediation, restorative activities, and reimbursement to habitats, biodiversity, and/or natural resources ability to operate and be protected within the law.

Ecological plasticity (or phenotypic plasticity) is the ability of an organism to modify its traits or behavior in light of changing environmental conditions. This means one genotype can produce many different phenotypes, enabling organisms to adapt and survive in a variety of or changing habitats.

Ecological threats are potential dangers to natural habitats and systems. These threats are frequently the result of human activities or natural disasters. Various ecological threats, such as climate change, can destabilize an ecosystem, resulting in loss of biodiversity and loss of materials.

Ecosystem dynamics refers to the flows of energy and nutrients, population fate, and the responses of an ecosystem to events and changes over time.

Endozoochorous dispersal is one type of seed dispersal process in which a seed is consumed by an animal and passes through the animal to be egested, either dead or viable in feces. Endozoochorous dispersal of seeds represents a mutualistic relationship and allows plants to have their offspring seeded away from the parent plant, sometimes even aiding in the seed germination process by passing through the animal gut.

Ethnobotanical heritage includes the body of knowledge, practices, and beliefs that link communities to plants. This heritage is a living expression of knowledge that has been passed down from generation to generation and includes traditional uses of plants for food, medicine, materials, and cultural practices.

Frugivorous birds are defined as those birds that consume mainly or entirely fruit. These birds are an important full or part of an ecosystem, providing seed dispersal. Similarly, frugivorous birds have benefits to plant dispersal and establishment and maintain forest planet /plant health.

Hardiness In biology, "hardiness" is defined as an organism's ability to survive and thrive in adverse environmental conditions. Specifies a plant's ability to resist the environmental stress of temperature extremes, drought, disease, and so forth.

Hepatotoxicity is the damage to an individual's liver caused by exposure to harmful substances, which may be medications, chemicals, or toxins. Liver injury may diminish the liver's capacity to function, ranging from mild and asymptomatic to severe cases of liver failure.

Invasion corridors are routes that allow invasive species (through human policies and processes) to expand into new habitats (e.g., canals, roads, disturbed areas). Natural corridors allow native species to expand into new habitats, whereas invasion corridors create pathways for nonnative species to invade (intentionally or unintentionally/local or distant) new habitats and negatively affect ecosystems.

Invasion ecology analyzes nonnative species, how they are introduced into new environments, become established, and spread. It investigates the ecological, evolutionary, and socioeconomic effects of these "invasive species" on native ecosystems.

Invasive species are nonindigenous organisms that have been introduced to a new area. Invasive species create harm, and they can cause ecological or economic harm created by animals, plants, or pathogens. Invasion species outcompete native species for resources and disrupt ecosystems, causing biodiversity loss and extinctions.

Morphology is a branch of biology that is concerned with the study of the forms and structures of organisms and their specific structural features. Morphology includes external morphology (the outward form), along with internal morphology (internal arrangement of parts or anatomy).

Niche displacement is similar to character displacement and involves competing species evolving to exploit different resources or habitats, and thus, they avoid (to the degree possible) direct competition and coexist by basically shifting what ecological role they perform.

Nutrient lift is when nutrients used by plants are reallocated from deeper soil layers or aquarium sediments to the surface or upper water column. (e.g., deep-rooted plants, bioturbation, upwelling). Nutrient lift makes the nutrients available for uptake by organisms.

Pharmacopoeia is an official and legally authoritative reference book providing standards of quality, purity, and strength of medicinal drugs. It describes in detail various pharmaceutical substances, their formulations, and tests for evaluating their quality, safety, and efficacy.

Photosensitization: caused by *Lantana camara,* only causes harm to livestock that ingest the plant. The herb has toxins (lantadenes) that harm the liver, preventing the excretion of phylloerythrin, a breakdown substance of chlorophyll. The accumulation of phylloerythrin in skin is sufficient that when introduced to sunlight, it causes severe skin inflammation, lesions, and necrosis and is amplified in unpigmented skin.

Phytochemistry is the study of the biochemicals derived from plants, focusing on structure, biosynthesis, and function. Phytochemistry also spans a broad range of plant compounds, including natural products for use in medicine, agriculture, and many other uses.

Phytoremediation is a technology that uses plants to clean up or restore contaminated soil, water, or air. Plants can absorb, degrade, or stabilize pollutants and are an economically viable solution to return an area to health and vitality.

Soil erosion is the removal of the top layer of soil, either by wind, water, or human reconstruction of land. This process causes significant land degradation, reduces agricultural yield, and can lead to desertification and siltation of water bodies.

Sustainable resource management is using the natural resource at a rate and in a manner that ensures its availability for future generations. This sustainably uses natural resources so that economic development occurs in balance with social equity and environmental protection, preventing resource depletion and degradation.

Taxonomy is the study of classifying organisms into taxonomic groups with a hierarchy based on shared traits. Taxonomy includes identifying, naming, and arranging species to reflect evolutionary relationships.

Toxicology is the scientific study of the adverse effects of chemical, biological, and physical agents on living organisms. It studies how a substance can cause harm, dose response, and how to prevent or treat harm.

Index